Chemical Reaction Engineering
Engineering
A First Course

Ian S. Metcalfe

Department of Chemical Engineering, University of Edinburgh.
Formerly Department of Chemical Engineering and Chemical Technology
Imperial College of Science, Technology, and Medicine
London

Series sponsor: **ZENECA**

ZENECA is a major international company active in four main areas of business: Pharmaceuticals, Agrochemicals and Seeds, Specialty Chemicals, and Biological Products.

ZENECA's skill and innovative ideas in organic chemistry and bioscience create products and services which improve the world's health, nutrition, environment, and quality of life. ZENECA is committed to the support of education in chemistry and chemical engineering.

OXFORD   NEW YORK   TOKYO
OXFORD UNIVERSITY PRESS
1997

Oxford University Press, Great Clarendon Street, Oxford OX2 6DP

Oxford   New York
Athens   Auckland   Bangkok   Bogota   Bombay   Buenos Aires
Calcutta   Cape Town   Dar es Salaam   Delhi   Florence   Hong Kong
Istanbul   Karachi   Kuala Lumpur   Madras   Madrid   Melbourne
Mexico City   Nairobi   Paris   Singapore   Taipei   Tokyo   Toronto   Warsaw

and associated companies in
Berlin   Ibadan

Oxford is a trade mark of Oxford University Press

Published in the United States
by Oxford University Press Inc., New York

A catalogue record for this book is available from the British Library

Library of Congress Cataloging in Publication Data
(Data applied for)

ISBN 0 19 8565380 (p/b)

Typeset by EXPO Holdings, Malaysia

Printed in Great Britain by Bath Press, Bath

Foreword

Lynn F. Gladden
University of Cambridge, Department of Chemical Engineering

Understanding how chemical reactors work lies at the heart of almost every chemical processing operation. This particular topic more than any other represents a true integration of the skills of the chemist and the chemical engineer.

Oxford Chemistry Primers are designed to give a concise introduction to a wide range of topics that may be encountered by chemistry and chemical engineering students. Each primer typically contains material that would be covered in an 8–10 lecture course. In this primer Ian Metcalfe presents a systematic, and very student-friendly introduction to the basic principles of chemical reaction engineering for students of both chemistry and chemical engineering. Each topic is illustrated with a wealth of examples, making this primer not only an excellent self-teaching aid but also extremely useful to those teaching this subject.

Preface

The book is primarily intended for use by chemical engineering under-graduates to complement their first chemical reaction engineering course at university. It is also intended to introduce students of other disciplines (such as chemistry) to the principles of chemical reaction engineering. It should be possible for an undergraduate to learn all of the fundamental aspects of reaction engineering from the book, using it as a self-study guide. For this reason the focus of the book is very much on underlying principles and not on detail. Consequently, a large number of example problems have been used. These examples come in two forms.

Firstly there are small problems used throughout the book to illustrate important points. These problems are immediately followed by the solution. It is recommended that the reader tries to solve these problems before viewing the solution. However, these examples are mainly intended to put the concepts in context and it is not necessary, or advisable, that the reader feels he or she must be able to solve each one independently before proceeding. The second type of problem is found at the end of each main chapter and is based upon final exams and problem sheets. Again, it is intended that the reader attempts to solve these problems before viewing the solutions. However, more importantly, the solution to the problem should be understood; if the problem is found to be difficult, viewing the solution is recommended.

I would like to thank Professor R. Kandiyoti for his help, Dr. M. Sahibzada for reading the manuscript and for making such useful comments, and my wife, Alison, for her support and patience. I would also like to thank all of the students I have taught who have helped me refine the ideas presented here.

London   I.S.M.
March 1997

Contents

1 Introduction

1.1 Scope

The book will start by showing in Chapter 2 how material balances should be performed for the three fundamental reactor types used in reaction engineering, namely the plug flow reactor (PFR), the continuous stirred tank reactor (CSTR), and the perfectly mixed batch reactor. In Chapter 3 we will see how these material balances can be used to design reactors (by this we primarily mean how to calculate their volume or residence time) when one reaction is taking place. We will also compare the behaviour of the different reactors. In Chapter 4 we will proceed to look at how the design process must be modified when more than one reaction is occurring. In Chapter 5 we will see that reactors need not be isothermal. Therefore, we need to look at how reaction rate depends upon temperature for different classes of reaction. Then we will formulate the energy balance for given reactors and use this to investigate the variation of temperature and therefore reaction rate with time or position in the reactor. This in turn will be used to allow us to calculate reactor volumes and residence times for a given duty. Finally, in Chapter 6, there is a brief discussion of non-ideal reactors; this is intended to illustrate the limitations of always assuming that reactors behave in an ideal manner.

Semi-batch or semi-continuous processes will not be considered. Likewise, the book will not discuss catalysis, mass transfer, or any related phenomena, but will lead into a second course covering such material. All reactions are presented as being homogeneous reactions and reaction rates are always presented as being volume specific (as opposed to being specific to the mass of catalyst used). The gas phase will always be treated as ideal.

1.2 Starting knowledge required

It is expected that all readers will have a British A-level or equivalent knowledge of mathematics such that they are comfortable with the use of differential calculus in applied situations and that they can perform simple analytical integration and numerical integration. They should be familiar with the concepts of material and energy balances and should be able to apply these in unfamiliar circumstances. This book is primarily intended for readers who have already had a basic course in chemical kinetics. As some readers may not have come across reaction kinetics before, there is a short description below, which, although far from complete, will allow the student to proceed with this text until they do take a course on chemical kinetics. The reader should also have some knowledge of thermodynamics for an appreciation of reaction equilibria, heats of reaction, and energy balances.

1.3 The rate expression

When material and energy balances for batch, CSTR, and PFR reactors are performed, reaction rate expressions will be required in an algebraic form.

Reaction rate is usually expressed in terms of the concentrations or partial pressures of the reactants (and sometimes products) and may be determined empirically or may, in part, be based upon an understanding of the reaction mechanism.

Consider the irreversible reaction,

$$\nu_A A \rightarrow \nu_B B$$

where ν_A and ν_B are stoichiometric coefficients. The rate of disappearance of A, r_A, can be given by a reaction rate expression depending upon the concentration of A,

$$r_A = k_C C_A^n$$

where n is the reaction order (not necessarily equal to ν_A), k_C is a rate constant, and C_A is the concentration of A[1].

Alternatively, the reaction could be expressed in terms of partial pressures (this is very common in the case of gas-phase reactions),

$$r_A = k_P P_A^n$$

where k_P is a rate constant and P_A is the partial pressure of A.

For a gas-phase reaction, k_P and k_C can be related,

$$k_C C_A^n = k_P P_A^n$$

For an ideal gas,

$$P_A V = N_A RT$$

where V is the volume of the batch reactor, N_A is the total number of moles of A in the reactor, R is the gas constant, and T is the reactor temperature (which must of course be in Kelvin).

$$P_A = \frac{N_A}{V} RT = C_A RT$$

where C_A is the concentration of A.

$$k_C C_A^n = k_P C_A^n (RT)^n$$
$$k_p (RT)^n = k_C$$

This relationship between k_P and k_C is true for all reactor types.

Consider a general reversible reaction,

$$\nu_A A + \nu_B B \Leftrightarrow \nu_C C + \nu_D D$$
$$r_{A1} = k_1 C_A^{n_A} C_B^{n_B}$$

where r_{A1} is the rate of forward reaction in terms of moles of A disappearing, n_A and n_B are reaction orders in those species, and,

$$r_{A-1} = k_{-1} C_C^{n_C} C_D^{n_D}$$

where r_{A-1} is the rate of the back reaction in terms of moles of A being formed. Therefore the net rate of disappearance of A is,

$$r_A = r_{A1} - r_{A-1}$$
$$r_A = k_1 C_A^{n_A} C_B^{n_B} - k_{-1} C_C^{n_C} C_D^{n_D}$$

At equilibrium the net rate of production or removal of A is equal to zero,

$$0 = k_1 C_A^{n_A} C_B^{n_B} - k_{-1} C_C^{n_C} C_D^{n_D}$$

[1] Throughout this text rates of reaction will be given in terms of one of the species participating in the reaction, hence the rates for other species may be calculated using stoichiometry, for example:

$$r_B = \frac{\nu_B}{\nu_A} r_A$$

where r_B is the rate of appearance of B. As regards the sign convention for the rate of reaction, in general, the rate of disappearance of a reactant will be defined as a positive rate and the rate of appearance of a product will be defined as positive. There are of course other ways of handling the sign convention; it is most important that whatever convention is used is applied consistently.

$$k_1 C_A^{n_A} C_B^{n_B} = k_{-1} C_C^{n_C} C_D^{n_D}$$

$$\frac{k_1}{k_{-1}} = K = \frac{C_C^{n_C} C_D^{n_D}}{C_A^{n_A} C_B^{n_B}}$$

where K is the equilibrium constant.

1.4 Objectives

This book has a number of different kinds of objectives. First of all there is a knowledge of the definitions and principles within reaction engineering that the reader is expected to gain while reading the book. Secondly, there are a number of skills that the reader should acquire. All of these skills combined will enable the reader to design basic chemical reactors (each one of these skills is graded to illustrate how it relates to the overall understanding of the reader). Thirdly, the reader's attitude towards reaction engineering (and in a larger sense engineering in general) should change by reading the book.

1.4.1 Knowledge

At the end of the book the reader should know

(a) the difference between batch, semi-batch, and continuous modes of operation;
(b) the principles of perfect mixing and plug flow and how they relate to the three basic reactor models used in the book—the perfectly mixed batch reactor, the continuous stirred tank reactor (CSTR), and the plug flow reactor (PFR);
(c) the definitions of space time and space velocity;
(d) that an infinite number of CSTRs in series behave like a PFR and that a PFR with infinite recycle rate behaves like a CSTR;
(e) that, for a first-order reaction a PFR will give a greater conversion than a CSTR because of the higher effective concentration of reactants;
(f) the definitions of selectivity and yield when multiple reactions occur;
(g) qualitatively, how the concentrations of reaction intermediates in a series reaction may vary with reactor residence time;
(h) in what manner reaction rates are influenced by changing temperature and composition;
(i) that there exists an optimum temperature for a reversible exothermic reaction where rate is maximized;
(j) what 'interstage cooling' and 'cold shot' are and when they are used.

1.4.2 Skills

At the end of the book the reader should be able to

(a) perform mass balances for the basic reactor models so as to derive the appropriate design equation from first principles *;
(b) apply the basic reactor models for the design of isothermal reactors given any set of reaction kinetics*;
(c) perform energy balances for the basic reactor models as necessary*;

(d)   apply energy balances for the design of reactors given any set of reaction kinetics*;

(e)   apply basic reactor models for the design of isothermal reactors when multiple reactions are taking place**;

(f)   in the case of multiple reactions, find the optimum reactor type and residence time for given kinetic behaviour and economic constraints**;

(g)   for non-isothermal reactors, maximize reaction rates by an appropriate choice of operating temperature, as necessary**;

(h)   design more complex reactor networks, e.g. reactors in series and parallel**;

(i)   generate more complex reactor models as necessary***;

(j)   critically evaluate the applicability and limitations of any reactor model for a given purpose***.

*Constitutes a basic understanding of the material covered.
**A more advanced understanding.
***A thorough understanding and appreciation of the material covered with the ability to apply this in unfamiliar circumstances.

1.4.3   Attitudes

At the end of the book the reader's attitude should move away from:
'All reactors are too complex to model' or 'All reactors are either perfectly mixed or plug flow',
and towards:
'It is possible to describe chemical reactors using models. However, all models have limitations. It is an engineer's responsibility to be aware of these limitations and to choose a model which is of sufficient complexity to give an answer of the required accuracy.'

2 Material balances for chemical reactors

The material (or mass balance) for a reactant can be written in a general form applicable to any type of reactor. Consider a small volume element of a reactor and what happens over a small interval of time. The material balance for any individual reactant or product is,

accumulation of = moles entering − moles leaving − moles disappearing (2.1)
moles in element   element per   element per   due to reaction per
per unit time   unit time   unit time   unit time
(1)   (2)   (3)   (4)

In general, some of these terms will be equal to zero, e.g. for batch reactors terms 2 and 3 are zero (if the element is the whole reactor), for a steady-state process term 1 is zero. We will apply this equation to the three types of basic reactor illustrated in Fig. 2.1.

In a batch reactor, see Fig. 2.1(a), all of the reactants are supplied to the reactor at the outset. The reactor is then sealed and the reaction is performed. There is no addition of reactants or removal of products during the reaction. The vessel is kept perfectly mixed so that there are no concentration or temperature gradients.

A plug flow reactor (PFR), see Fig. 2.1(b), is a special type of tubular reactor. Feed is continuously supplied to the reactor and products are continually removed. There is no attempt to induce mixing in the reactor. The velocity profile is flat, i.e. uniform over any cross-section normal to the direction of fluid motion.

The continuous stirred tank reactor (CSTR), see Fig. 2.1(c), like the PFR, has a continuous supply of feed while products are continually removed. However, in this case perfect mixing is achieved, i.e. there are no concentration or temperature gradients within the reactor (in a similar manner to the batch reactor).

2.1 Batch reactors

As previously mentioned, the total feed is introduced at the outset and no withdrawal is made until the reaction has reached the degree of completion desired.

1. This is a fundamentally unsteady process. We expect all variables to change with time.
2. The mixture in the volume is assumed to be perfectly mixed. This means that there will be uniform concentrations.

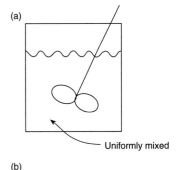

(a)

Uniformly mixed

(b)

Feed   Product

(c)

Feed

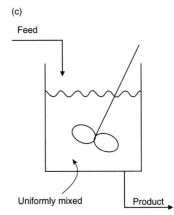

Uniformly mixed   Product

Fig. 2.1 Schematic representation of: (a) a batch reactor; (b) a plug flow reactor; (c) a continuous stirred tank reactor.

3. The temperature will also be uniform throughout the reactor—however, it may change with time.

4. The volume may be kept fixed (most common) or may be varied to keep the pressure constant (i.e. there are two modes of operation, constant volume or constant pressure).

First of all we must identify the element over which we will perform the material balance, then we can simplify our fundamental material balance (eqn 2.1). Because the concentrations are uniform within the reactor it is most convenient to treat the entire reactor as this element[1]. With this element, terms 2 and 3 in eqn 2.1 go to zero as there is no addition or removal of material. For reactant A in the volume V between times t and $t + \Delta t$ where,

N_A is number of moles of A in volume V (the system)

r_A is reaction rate defined as $\dfrac{\text{moles of A disappearing}}{\text{(unit time) (unit volume)}}$

the material accumulated (term 1) in the time interval is given by,

$$\text{moles accumulated} = N_A|_{t+\Delta t} - N_A|_t$$

and the material reacted in the same time interval in the reactor (term 4) is given by,

$$\text{moles reacted} = r_A V \Delta t$$

Therefore, using the material balance (eqn 2.1) we can substitute mathematical expressions for all of the terms,

$$N_A|_{t+\Delta t} - N_A|_t = 0 + 0 - r_A V \Delta t$$

$$\frac{N_A|_{t+\Delta t} - N_A|_t}{\Delta t} = -r_A V$$

In the limit as $\Delta t \to 0$,

$$\frac{dN_A}{dt} = -r_A V \quad \text{or}$$

$$r_A = -\frac{1}{V}\frac{dN_A}{dt} \tag{2.2}$$

Equation 2.2 is often known as the design equation for batch reactors. We will come back to this equation and apply it when we look at batch reactor design in more detail.

Note that we have defined the rate in terms of the disappearance of A, hence, a minus sign appears in the equation as dN_A/dt is negative (the amount of A in the reactor is decreasing with time). Furthermore, the greater the reaction rate, i.e. the larger r_A, the quicker A will disappear.

If the rate is defined in terms of the rate of production of a product B we can say,

r_B is reaction rate defined as $\dfrac{\text{moles of B formed}}{\text{(unit time) (unit volume)}}$

$$r_B = \frac{1}{V}\frac{dN_B}{dt} \tag{2.3}$$

This time there is no negative sign in the equation as the number of moles of B in the reactor will increase with time[2].

[1] In fact it would be very difficult to use any other element because we do not know enough about the flow-field within the reactor. A fuller explanation is complicated, however; the fluxes of any species into and out of a differential element in a perfectly mixed reactor will be very large but this does not mean that they are equal. In fact, this inequality of fluxes is needed to maintain concentration gradients at negligible levels—negligible in so far as one average concentration can be accurately used for the determination of the reaction rate.

[2] Once again, for sign conventions it is important to think about the physical meaning of the equation and to remain consistent.

The time necessary for achieving a given degree of reaction can be found by integrating the design equation with the initial condition, $t = 0$; $N_A = N_{A0}$

$$\int_0^t dt = -\int_{N_{A0}}^{N_A} \frac{1}{V} \frac{dN_A}{r_A} \tag{2.4}$$

Remember: the reaction will proceed (unless externally arrested) until: (a) equilibrium is reached (reversible reaction); (b) limiting reactant is exhausted (irreversible reaction).

The degree of reaction that is required in a reactor is often specified in terms of conversion, i.e. x_A being the conversion of A[3]. At any time, t,

$$x_A = \frac{N_{A0} - N_A}{N_{A0}} \tag{2.5}$$

In the limit of 0% conversion,

$$N_A = N_{A0}$$

while in the limit of 100% conversion,

$$N_A = 0$$

Differentiating eqn 2.5 we get,

$$dx_A = -\frac{dN_A}{N_{A0}} \tag{2.6}$$

and this can be used to change the variable of integration in eqn 2.4,

$$\int_0^t dt = \int_0^{x_A} \frac{N_{A0}}{V} \frac{dx_A}{r_A} \tag{2.7}$$

A first-order irreversible reaction,

$$A \rightarrow B; r_A = kC_A$$

is to be carried out in a batch reactor. If $k = 0.01 \text{ s}^{-1}$, calculate the time to reach 30% conversion.

From eqn 2.4,

$$t = -\int_{N_{A0}}^{N_A} \frac{1}{V} \frac{dN_A}{r_A}$$

$$A \rightarrow B; r_A = kC_A$$

The reaction rate is expressed in terms of concentration but we must put everything on the right-hand side of the equation in terms of one variable to integrate. This can be done by recalling that the concentration is simply the number of moles divided by the total volume.

$$r_A = kC_A = kN_A/V$$

$$t = -\int_{N_{A0}}^{N_A} \frac{dN_A}{kN_A} = -\frac{1}{k} \ln \frac{N_A}{N_{A0}}$$

---

[3] If we have more than one reactant, e.g.

$$A + B \rightarrow \text{Products}$$

then the conversion of A and B will not, in general, be the same. Mostly, but not always, we will work with the conversion of the limiting reactant.

We will not use the concept of extent of reaction in this text. The extent of reaction is defined as,

$$\xi = \frac{N_{i0} - N_i}{v_i}$$

where v_i is the stoichiometric coefficient of the ith species. No matter which species is considered, the extent of reaction will then be the same (unlike conversion which is species dependent). However, when working with extent of reaction, care must be taken to write the reaction in one consistent form so that the stoichiometric coefficients remain fixed.

Example 2.1

Solution

$$x_A = \frac{N_{A0} - N_A}{N_{A0}} = 1 - \frac{N_A}{N_{A0}}$$

$$t = -\frac{1}{k}\ln(1 - x_A) = 35.7 \text{ s}$$

2.2 Plug flow reactors (PFRs)

Now that we have 'designed' a simple batch reactor we will take a look at plug-flow reactors.

Tubular reactors are used for many large-scale gas reactions, e.g.

- Homogeneous reactions (tube contains only the reactant and product gases and any inert gases), e.g.

$$2NO + O_2 \rightarrow 2NO_2 \qquad (HNO_3 \text{ from } NH_3)$$

- Heterogeneous reactions (tube is packed with catalyst), e.g.

$$CO + 2H_2 \Leftrightarrow CH_3OH$$

$$N_2 + 3H_2 \Leftrightarrow 2NH_3$$

In tubular reactors, there is a steady movement of reagents in a chosen direction. No attempt is made to induce mixing of fluid between different points along the overall direction flow.

A formal material balance over a differential volume element requires that we know (or assume) patterns of fluid behaviour within the reactor, e.g. the velocity profile. The simplest set of assumptions about the fluid behaviour in a tubular reactor is known as the plug flow (or piston flow) assumption. Reactors approximately satisfying this assumption are called plug flow reactors (PFRs).

The plug flow assumptions are as follows,

(a) flow rate and fluid properties are uniform over any cross-section normal to fluid motion;

(b) there is negligible axial mixing—due to either diffusion or convection.

The plug flow assumptions tend to hold when there is good radial mixing (achieved at high flow rates $Re > 10^4$) and when axial mixing may be neglected (when the length divided by the diameter of the reactor ≥ 50 (approximately)).

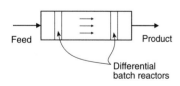

Feed — Product

Differential
batch reactors

Fig. 2.2 Schematic of a PFR showing how differential elements may be considered to be well-mixed batch reactors.

This means that if we consider a differential element within the reactor (with its boundaries normal to the fluid motion), it can be taken to be perfectly mixed and, as it travels along the reactor, it will not exchange any fluid with the element in front of or behind it. In this way, it may be considered to behave as a differential batch reactor (see Fig. 2.2). We will see later that there are consequently many similarities between the behaviour of a PFR and the behaviour of a batch reactor.

Furthermore, we will make an additional assumption that the reactor is at steady state. In a batch reactor, composition changes from moment to moment.

In continuous operation at steady state in a PFR, the composition changes with position, but at a given position there is no change with time.

We are now able to perform a material balance (we now know enough about the behaviour of the fluid within the reactor).

We must choose an appropriate element over which to perform the material balance. It might seem natural to follow one of these 'differential batch reactors', i.e. to perform the material balance over such an element and integrate over the time taken for the element to travel along the tube. However, this would lead to the same equation as the batch reactor and, because of its time dependency, would imply that the concentrations in the reactor were changing with time (this approach is not incorrect but we must remember that the time refers to the residence time of the fluid in the reactor). Alternatively, we choose to observe what happens in a differential element that is in a fixed position at any point along the length of the reactor (we do not follow elements—we watch them go by). See Fig. 2.3.

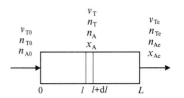

Fig. 2.3 Schematic of a PFR with nomenclature and showing a differential element used in the formulation of the material balance.

The following symbols are used[4],

l	length (m)
L	total reactor length (m)
V	reactor volume (m^3)
A	cross-sectional area (m^2)
n_A	molar flow rate of component A (mol s^{-1})
n_T	total molar flow rate (mol s^{-1})
v_T	volumetric flow rate (m^3 s^{-1})
0	denotes inlet conditions (subscript)
e	denotes exit conditions (subscript)

[4] A lower case n is now used to denote that we are considering molar flow rates (and not an absolute number of moles as in the case of the batch reactor when an upper case N was used).

The material balance over the differential element has the usual form,

Accumulation = input − output − loss through reaction

We now substitute in mathematical expressions for each of these terms. There is no accumulation in the element as the reactor is at steady state. Therefore, the element always contains fluid of the same composition. The input and the output from the element are just the molar flows of whatever component we are considering and the loss through reaction will just be the reaction rate (which is volume specific) multiplied by the volume of the element.

$$0 = n_A|_l - n_A|_{l+dl} - r_A dV$$

Dividing throughout by dl the differential is obtained,

$$0 = -\frac{dn_A}{dl} - r_A\frac{dV}{dl}$$

Dividing by the cross-sectional area and remembering that $Adl = dV$,

$$r_A = -\frac{dn_A}{dV} \tag{2.8}$$

Equation 2.8 is known as the design equation for PFRs. As can be seen, the higher the reaction rate the faster the molar flow of A will decrease as the reactor length or volume is increased. Whereas in the case of a batch reactor compositional changes take place with time, in a tubular reactor (or our special case of a PFR) the compositional changes take place spatially.

The design equation could also be written for a product of the reaction,

$$r_B = \frac{dn_B}{dV} \tag{2.9}$$

For greater reaction rates, the flow of B will increase even faster as we pass along the length of the reactor.

The dependence of the volume of a PFR on conversion may be found by performing the integration of the design equation (eqn 2.8),

$$\int_0^V dV = V = -\int_{n_{A0}}^{n_A} \frac{dn_A}{r_A} \tag{2.10}$$

The definition for conversion (eqn 2.5) can be rewritten in terms of molar flow rates,

$$x_A = \frac{n_{A0} - n_A}{n_{A0}} \tag{2.11}$$

Differentiating gives,

$$dx_A = -\frac{dn_A}{n_{A0}} \tag{2.12}$$

Expressing eqn 2.10 in terms of conversion of reactant we get,

$$V = \int_0^{x_A} \frac{n_{A0} dx_A}{r_A} \tag{2.13}$$

Example 2.2   A first-order irreversible reaction,

$$A \rightarrow B; \, r_A = kC_A$$

is to be carried out in a plug flow reactor. If $k = 0.01$ s^{-1} and the volumetric flow rate is 10^{-3} m^3 s^{-1}, calculate the reactor volume and residence time required for 30% conversion.

Solution   From eqn 2.10,

$$V = -\int_{n_{A0}}^{n_A} \frac{dn_A}{r_A}$$

$$A \rightarrow B; \, r_A = kC_A = kn_A/v_T$$

$$V = -\frac{v_T}{k} \int_{n_{A0}}^{n_A} \frac{dn_A}{n_A}$$

$$= -\frac{v_T}{k} \ln \frac{n_A}{n_{A0}}$$

$$= -\frac{v_T}{k} \ln(1 - x_A) = 3.57 \times 10^{-2} \text{ m}^3$$

Residence time, $\tau = V/v_T = 35.7$ s

If we compare the result from Example 2.2 to that from Example 2.1 we can see that the residence time required in both cases is the same. This is because we can consider the elements in the PFR to be differential batch reactors. Their composition–time history is just the same as if they were in a batch reactor. The difference between batch and PFR reactors is that composition changes take place temporally in the first and spatially in the second.

2.3 Continuous stirred tank reactors (CSTRs)

The key feature of a CSTR is that perfect mixing occurs in the reactor. Perfect mixing means that the properties of the reaction mixture are uniform in all parts of the vessel and identical to the properties of the reaction mixture in the exit stream. Furthermore, the inlet stream instantaneously mixes with the bulk of the reactor volume (see Fig. 2.4).

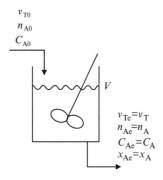

Fig. 2.4 Schematic of a CSTR with nomenclature.

We will also assume that the CSTR has reached steady state. Therefore reaction rate is the same at every point, and time independent.

Again we must choose an element for our material balance. Like the batch reactor it would be very difficult to consider a differential element because we do not know enough about the flow field in the reactor. Furthermore, because the properties of the reaction mixture are uniform throughout the reactor volume we can perform the material balance for component A over the whole volume, V.

Again,

Accumulation = input − output − loss through reaction

We now substitute in mathematical expressions for each of these terms. There is no accumulation in the element as the reactor is at steady state. The input and the output from the element are just the molar flows to and from the reactor and the loss through reaction will just be the reaction rate (which is volume specific) multiplied by the volume of the reactor.

$$0 = n_{A0} - n_{Ae} - r_A V$$
$$r_A = \frac{n_{A0} - n_{Ae}}{V} \tag{2.14}$$

This equation is known as the design equation for a CSTR.

The design equation can be rewritten because the conditions within the reactor are the same as those in the outlet stream and therefore there is no need to distinguish between them. More commonly, the design equation is simply written as,

$$r_A = \frac{n_{A0} - n_A}{V} \tag{2.15}$$

The design equation for a CSTR is not a differential equation as it is in the case of a PFR or batch reactor. The PFR needs a differential equation to describe it because compositional changes take place spatially and the batch reactor obeys a differential equation because compositional changes take place in time. The equation for a CSTR, in contrast, is not a differential equation because composition does not change spatially or temporally. So we could ask, if there are no changes of composition spatially or temporally how can the CSTR behave as a reactor? The answer lies in the fact that there is a discontinuity where fresh feed is introduced into the reactor. Because of perfect mixing there is an instantaneous change in the composition and it is here that the compositional change takes place.

Example 2.3   A first-order irreversible reaction,

$$A \rightarrow B; r_A = kC_A$$

is to be carried out in a CSTR. If $k = 0.01 \text{ s}^{-1}$ and the volumetric flow rate is $10^{-3} \text{ m}^3 \text{ s}^{-1}$, calculate the reactor volume and residence time required for 30% conversion. (Remember that the concentration in the reactor is the same as the outlet concentration because of perfect mixing.)

Solution   From eqn 2.15,

$$V = \frac{n_{A0} - n_A}{r_A}$$

$$A \rightarrow B; r_A = kC_A = kn_A/v_T$$

$$V = \frac{n_{A0} - n_A}{k \frac{n_A}{v_T}}$$

$$= \frac{v_T}{k}\left(\frac{n_{A0}}{n_A} - 1\right)$$

But $x_A = 1 - \dfrac{n_A}{n_{A0}}$

$$V = \frac{v_T}{k}\left[\frac{1}{(1 - x_A)} - 1\right] = \frac{v_T}{k}\frac{x_A}{(1 - x_A)} = 4.29 \times 10^{-2} \text{ m}^3$$

Mean residence time, $\tau = V/v_T = 42.9 \text{ s}$

As Example 2.3 shows, the volume or residence time required for a given conversion in a CSTR (for a first-order reaction) is greater than that required in a PFR or batch reactor (compare with Examples 2.1 and 2.2). This is because when we introduce fresh feed into the CSTR we immediately dilute it to the exit concentration and therefore we see lower rates and hence longer residence times are required. In a PFR the concentration of reactants gradually reduces as we travel along the length of the reactor so, although the rate at the exit of the PFR is similar to that in the CSTR, the upstream rates are higher. In a batch reactor at short times we have higher rates and the rate gradually decreases as time passes.

Imagine having a reaction taking place in the liquid phase where the reactant is red and the product is colourless. The feed to the reactor will be red. In the case of a CSTR, the exit stream will be a light shade of red. This will be the same colour as the liquid within the reactor. In the case of a PFR the exit stream will also be a light shade of red; however, the colour of the stream will change gradually from inlet to outlet. For reactions of positive order in concentration, deeper shades of red would of course correspond to higher reaction rates.

However, not all rates are higher at higher reactant concentrations. We could be designing a reactor for a reaction which obeyed kinetics in which the rate depended, for instance, on the inverse of a reactant concentration. In this case lower concentrations would give higher rates and we would find therefore that a CSTR would require a smaller volume or residence time than a PFR or batch reactor for the same conversion. This comparison between CSTR and PFRs is a subject that we will return to and investigate in more detail in the next chapter.

3 Calculation of reactor volume and residence time

We will now apply the material balances or design equations derived in Chapter 2 to the design of isothermal reactors with one reaction occurring.

3.1 Residence time of batch reactors

Recalling the design equation (eqn 2.2)

$$r_A = -\frac{1}{V}\frac{dN_A}{dt}$$

The initial condition is that at time $t = 0$, $N_A = N_{A0}$
  Integration gives eqn 2.4,

$$\int_0^t dt = -\int_{N_{AO}}^{N_A} \frac{1}{V}\frac{dN_A}{r_A}$$

In general, for PFR and batch reactors, where we need to perform an integration, we will need to put everything on the right-hand side of the equation in terms of one common variable. This common variable will usually be total conversion[1]. Recalling the definition of conversion (eqn 2.5),

$$x_A = \frac{N_{A0} - N_A}{N_{A0}}$$

Differentiating eqn 2.5 gives eqn 2.6,

$$dx_A = -\frac{dN_A}{N_{A0}}$$

and eqn 2.6 can be used to change the variable of integration in eqn 2.4 yielding eqn 2.7,

$$\int_0^t dt = \int_0^{x_A} \frac{N_{A0}}{V}\frac{dx_A}{r_A}$$

In general, the irreversible reaction:

$$A \rightarrow \text{Products}$$

has a rate expression of the form

$$r_A = kC_A^n$$

Expressing this in terms of conversion,

$$r_A = kC_A^n = k\frac{N_A^n}{V^n} = k\frac{N_{A0}^n(1 - x_A)^n}{V^n}$$

This can be substituted into eqn 2.7,

$$t = \frac{1}{kN_{A0}^{n-1}} \int_0^{x_A} V^{n-1} \frac{dx_A}{(1 - x_A)^n} \tag{3.1}$$

N_{A0} is a constant and k is a constant (we assume that the reactor temperature does not change as the conversion changes).

There will be a change in the total number of moles present if the number of moles of products is different from the number of moles of reactants. For a gas-phase reaction, this means that, if the reactor is at constant pressure, the reactor volume may change (like a piston) or if the reactor is of constant volume the reactor pressure may change[2]. To proceed we need to know which type of reactor to consider.

[2] When reactions take place in the liquid phase between dissolved species there is no volume change or pressure change.

3.1.1   Constant volume batch reactor

The reactor volume is constant and so can be taken out of the integration (eqn 3.1):

$$t = \frac{V^{n-1}}{kN_{A0}^{n-1}} \int_0^{x_A} \frac{dx_A}{(1 - x_A)^n}$$

$$t = \frac{1}{kC_{A0}^{n-1}} \int_0^{x_A} \frac{dx_A}{(1 - x_A)^n}$$

and this expression can be easily integrated.

If $n \neq 1$,

$$t = \frac{-1}{kC_{A0}^{n-1}(1 - n)} [(1 - x_A)^{1-n}]_0^{x_A}$$

$$t = \frac{-1}{kC_{A0}^{n-1}(1 - n)} \{[(1 - x_A)^{1-n} - 1\}$$

For $n = 1$,

$$t = \frac{1}{k} \int_0^{x_A} \frac{dx_A}{(1 - x_A)} = -\frac{1}{k} \left[\ln(1 - x_A) \right]_0^{x_A}$$

$$t = \frac{1}{k} \ln(1 - x_A) = \frac{1}{k} \ln \frac{1}{(1 - x_A)} \quad \text{and} \tag{3.2}$$

$$x_A = 1 - e^{-kt} \tag{3.3}$$

This means that, for a first-order irreversible reaction, as the residence time of a batch reactor approaches infinity the conversion of the reaction will approach unity in an exponential manner.

Alternatively, expressing eqns 3.2 and 3.3 in terms of the moles of A present,

$$t = \frac{1}{k} \ln \frac{N_{A0}}{N_A} \quad \text{and} \tag{3.4}$$

$$N_A = N_{A0} e^{-kt} \tag{3.5}$$

3.1.2   Constant pressure batch reactor

Recalling eqn 3.1,

$$t = \frac{1}{kN_{A0}^{n-1}} \int_0^{x_A} V^{n-1} \frac{dx_A}{(1 - x_A)^n}$$

For a constant pressure batch reactor, the reactor volume is no longer constant and must remain within the integration. To perform the integration a relationship between the reactor volume and the conversion must be found.

Consider a general reaction,

$$\nu_A A \rightarrow \nu_B B + \nu_C C$$

where all of the reactants and products are gas-phase species. Let us consider the number of moles of each component present. First of all, from the definition of conversion (eqn 2.5), the number of moles of A at any time is given by,

$$N_A = N_{A0} - N_{A0} x_A$$

If we know how many moles of A have disappeared then we can calculate from the stoichiometry of the reaction how much B and C must have been produced,

$$N_B = N_{B0} + \frac{\nu_B}{\nu_A} N_{A0} x_A$$

$$N_C = N_{C0} + \frac{\nu_C}{\nu_A} N_{A0} x_A$$

The number of moles of inert present will remain constant,

$$N_I = N_{I0}$$

And if we sum over all of the components present we obtain an expression for the total number of moles present in the reactor.

$$N_T = N_{T0} + \left(\frac{\nu_B + \nu_C - \nu_A}{\nu_A} \right) N_{A0} x_A$$

$$\frac{N_T}{N_{T0}} = 1 + \left(\frac{\nu_B + \nu_C - \nu_A}{\nu_A} \right) \frac{N_{A0}}{N_{T0}} x_A = 1 + \varepsilon_A x_A$$

where ε_A is a constant that depends upon the stoichiometry and the feed conditions,

$$\varepsilon_A = \left(\frac{\nu_B + \nu_C - \nu_A}{\nu_A} \right) \frac{N_{A0}}{N_{T0}}$$

If we are dealing with an ideal gas,

$$PV = N_T RT; \quad PV_0 = N_{T0} RT$$

The volume is simply proportional to the number of moles present (at constant temperature and pressure),

$$\frac{V}{V_0} = \frac{N_T}{N_{T0}} = 1 + \varepsilon_A x_A$$

$$V = V_0 (1 + \varepsilon_A x_A) \qquad (3.6)$$

Equation 3.6 can now be substituted into the integration (eqn 3.1),

$$t = \frac{V_0^{n-1}}{k N_{A0}^{n-1}} \int_0^{x_A} (1 + \varepsilon_A x_A)^{n-1} \frac{dx_A}{(1 - x_A)^n}$$

$$= \frac{1}{k C_{A0}^{n-1}} \int_0^{x_A} \frac{(1 + \varepsilon_A x_A)^{n-1}}{(1 - x_A)^n} dx_A$$

Example 3.1   Consider the gas-phase reaction,

$$2A \rightarrow B + C + 3D \quad \text{with } r_A = kC_A^2$$

Initially 50% A and 50% inert are present. If the pressure and temperature remain constant, derive an expression relating the volume of the system to its initial volume and the conversion of A. (Assume ideal gas behaviour.)

Solution

$$N_A = N_{A0} - N_{A0}x_A, \quad N_B = N_{B0} + \frac{1}{2}N_{A0}x_A$$

$$N_C = N_{C0} + \frac{1}{2}N_{A0}x_A, \quad N_D = N_{D0} + \frac{3}{2}N_{A0}x_A$$

$$N_I = N_{I0}$$

$$N_T = N_{T0} + \frac{3}{2}N_{A0}x_A$$

$$\frac{N_T}{N_{T0}} = 1 + \frac{3}{2}\frac{N_{A0}}{N_{T0}}x_A = 1 + \frac{3}{4}x_A$$

$$PV = N_T RT; \quad PV_0 = N_{T0}RT$$

$$\frac{V}{V_0} = \frac{N_T}{N_{T0}} = 1 + \frac{3}{4}x_A$$

$$V = V_0\left(1 + \frac{3}{4}x_A\right)$$

3.2   PFRs

Again we take the design equation (eqn 2.8),

$$r_A = -\frac{dn_A}{dV}$$

and integrate it to give eqn 2.10,

$$V = -\int_{n_{A0}}^{n_A} \frac{dn_A}{r_A}$$

Putting everything in terms of one common variable, x_A, by using eqn 2.12,

$$dn_A = -n_{A0}dx_A$$

and substituting for nth-order kinetics,

$$r_A = kC_A^n = k\frac{n_A^n}{v_T^n} = k\frac{n_{A0}^n(1 - x_A)^n}{v_T^n} \quad \text{gives,}$$

$$V = \int_0^{x_A} \frac{v_T^n dx_A}{kn_{A0}^{n-1}(1 - x_A)^n} \tag{3.7}$$

However, the volumetric flow rate, v_T, need not be a constant (it will be constant for a liquid-phase reaction or for a gas-phase reaction in which the total number of moles and the pressure do not change). If it is not constant, we need an expression relating the volumetric flow rate to the conversion. Again for a general reaction,

$$\nu_A A \rightarrow \nu_B B + \nu_C C$$

where all of the reactants and products are gas-phase species, we can look at how the individual molar flow rates will depend upon the conversion (in a similar way to that of the batch reactor).

$$n_A = n_{A0} - n_{A0}x_A$$

$$n = n_{B0} + \frac{\nu_B}{\nu_A} n_{A0}x_A$$

$$n_C = n_{C0} + \frac{\nu_C}{\nu_A} n_{A0}x_A$$

$$n_I = n_{I0}$$

$$n_T = n_{T0} + \left(\frac{\nu_B + \nu_C - \nu_A}{\nu_A}\right) n_{A0}x_A$$

$$\frac{n_T}{n_{T0}} = 1 + \left(\frac{\nu_B + \nu_C - \nu_A}{\nu_A}\right)\frac{n_{A0}}{n_{T0}}x_A = 1 + \varepsilon_A x_A$$

Ideal gas behaviour,

$$P\upsilon_T = n_T RT; \quad P\upsilon_{T0} = n_{T0}RT$$

and we will assume that the pressure is equal everywhere within the reactor. (This is generally a good approximation but is not always true, particularly if the reactor contains a catalyst.) Therefore, the volumetric flow rate is proportional to the total molar flow rate[3],

$$\frac{\upsilon_T}{\upsilon_{T0}} = \frac{n_T}{n_{T0}} = 1 + \varepsilon_A x_A$$

$$\upsilon_T + \upsilon_{T0}(1 + \varepsilon_A x_A) \tag{3.8}$$

And substituting into eqn 3.7,

$$\upsilon = \frac{\upsilon_{T0}^n}{kn_{A0}^{n-1}} \int_0^{x_A} \frac{(1 + \varepsilon_A x_A)^n}{(1 - x_A)^n} \, dx_A$$

This equation is true for any *n*th-order irreversible reaction occurring in a PFR.

If we take a look at one special case, a reaction for which the volumetric flow rate is constant (so $\varepsilon_A = 0$ and $\upsilon_T = \upsilon_{T0}$) and which obeys first-order kinetics ($n = 1$), then the reactor volume as a function of x_A is given by,

$$V = \frac{\upsilon_T}{k} \int_0^{x_A} \frac{1}{(1 - x_A)} \, dx_A$$

$$V = -\frac{\upsilon_T}{k}[\ln(1 - x_A)]_0^{x_A} = \frac{\upsilon_T}{k} \ln \frac{1}{(1 - x_A)} \quad \text{or} \tag{3.9}$$

$$\frac{1}{(1 - x_A)} = \exp\left(\frac{kV}{\upsilon_T}\right) = \exp(k\tau)$$

where τ is the residence time of the reactor (this is equal to the volume of the reactor divided by the volumetric flow rate)[4]. Therefore,

$$x_A = 1 - e^{-k\tau} \tag{3.10}$$

[3] Compare eqn 3.8 with eqn 3.6.

[4] Question: if the volumetric flow rate changes as we proceed along the length of a reactor, how do we know what the residence time will be?

Answer: we need to perform an integration. If we consider a differential element we know that the residence time in this element will be dV/υ_T. Summing the residence times of all such elements,

$$\text{Total residence time} = \int \frac{dV}{\upsilon_T}$$

For real reactors the residence time can, consequently, be very difficult to calculate. As a result, what is known as the 'space-time' is often used. This space-time is equal to what the residence time would be if the volumetric flow rate remained unchanged at its original inlet value.

$$\text{Space-time} = \frac{V}{\upsilon_{T0}}$$

The term 'space-velocity', S_v, is also used. This is simply equal to the reciprocal of space-time,

$$S_v = \frac{\upsilon_{T0}}{V}$$

This means that in a PFR (as in a batch reactor, eqn 3.3) the conversion will approach unity exponentially as residence time is increased.

Expressing eqns 3.9 and 3.10 in terms of the molar flow of reactant, we get, for a first-order reaction with no volume change,

$$V = \frac{v_T}{k} \ln \frac{n_{A0}}{n_A} \quad \text{or,} \tag{3.11}$$

$$n_A = n_{A0} e^{-k\tau} \tag{3.12}$$

Example 3.2 For gas-phase reactions, rates are often expressed in terms of partial pressures,

e.g $\quad A \rightarrow B + 2C; \quad r_A = kP_A, \quad k = 50 \text{ mol s}^{-1} \text{ m}^{-3} \text{ bar}^{-1}$

In a PFR, the feed is pure A at a flow rate of 10 mol s^{-1} and the reactor operates at a pressure of 20 bar. Calculate the reactor volume required to reach 50% conversion.

Solution

$$V = -\int_{n_{A0}}^{n_A} \frac{dn_A}{r_A} = -\int_{n_{A0}}^{n_A} \frac{dn_A}{kP_A} = \int_0^{x_A} \frac{n_{A0} dx_A}{kP_A}$$

$$n_A = n_{A0} - n_{A0} x_A$$

$$n_B = n_{A0} x_A$$

$$n_C = 2 n_{A0} x_A$$

$$n_T = n_{T0} + 2 n_{A0} x_A$$

$$P_A = \frac{n_A}{n_T} P = \frac{n_{A0}(1 - x_A)}{n_{A0}(1 + 2x_A)} P = \frac{1 - x_A}{1 + 2x_A} \cdot P$$

$$V = \int_0^{x_A} \frac{n_{A0}(1 + 2x_A)}{k(1 - x_A)P} dx_A$$

$$= \frac{n_{A0}}{kP} \int_0^{x_A} (-2 + \frac{3}{(1 - x_A)}) dx_A$$

$$= \frac{n_{A0}}{kP} [-2x_A - 3 \ln(1 - x_A)]_0^{x_A}$$

$$= \frac{n_{A0}}{kP} (-2x_A - 3 \ln(1 - x_A))$$

$$= \frac{10}{(50)(20)} (1.079) \text{ m}^3 = 1.08 \times 10^{-2} \text{ m}^3$$

Example 3.3 Calculate the reactor volume required to achieve 90% conversion in an isothermal PFR for the second-order irreversible reaction,

$$2A \rightarrow B + C$$

The feed consists of pure A at a molar flow rate of 1.2 mol s^{-1}. The reaction rate constant is 6.7 mol s^{-1} m^{-3} bar^{-2}. The pressure of the reactor is 1.4 bar at the inlet, falling linearly with reactor length, to 1.0 bar at the outlet. The reactor is of uniform cross-sectional area.

$$r_A = -\frac{dn_A}{dV}, \quad r_A = kP_A^2$$

$$n_A = n_{A0}(1 - x_A), \quad n_{A0} = n_T, \quad P_A = \frac{n_A}{n_T}P$$

$$r_A = k\left\{\frac{n_{A0}(1 - x_A)}{n_{A0}}\right\}^2 P^2$$

$$\frac{-dn_A}{dV} = n_{A0}\frac{dx_A}{dV} = k\left\{\frac{n_{A0}(1 - x_A)}{n_{A0}}\right\}^2 P^2$$

$$\int_0^V dV = \int_0^{x_A} \frac{n_{A0}dx_A}{k(1 - x_A)^2 P^2}$$

The above equation cannot be integrated as the pressure cannot be expressed in terms of the conversion. However, the pressure can be expressed in terms of length which can be related to reactor volume.

$$\int_0^V P^2 dV = \int_0^{x_A} \frac{n_{A0}dx_A}{k(1 - x_A)^2}$$

$$P = P_0 - \frac{l}{L}\Delta P$$

where l is the length variable and L is the total reactor length.

$$\int_0^V \left(P_0 - \frac{l}{L}\Delta P\right)^2 dV = \frac{n_{A0}}{k}\int_0^{x_A} \frac{dx_A}{(1 - x_A)^2}$$

Using the length variable in the integration, $dV = Adl$,

$$A\int_0^L \left(P_0 - \frac{l}{L}\Delta P\right)^2 dl = \frac{n_{A0}}{k}\int_0^{x_{Ae}} \frac{dx_A}{(1 - x_A)^2}$$

(for the limits of integration remember that, formally, when $l = L$, $x_A = x_{Ae}$)

$$A\left[\frac{1}{3}\left(P_0 - \frac{l}{L}\Delta P\right)^3\left(\frac{-L}{\Delta P}\right)\right]_0^L = \frac{n_{A0}}{k}[(1 - x_A)^{-1}]_0^{x_{Ae}}$$

$$-\frac{AL}{3\Delta P}[(P_0 - \Delta P)^3 - (P_0)^3] = \frac{n_{A0}}{k}\left[\frac{1}{(1 - x_{Ae})} - 1\right]$$

$$AL = V = \frac{3\Delta P}{P_0^3 - P_e^3}\frac{n_{A0}}{k}\frac{x_{Ae}}{(1 - x_{Ae})} = \frac{3(0.4)}{(1.4)^3 - (1.0)^3}\frac{(1.2)(0.9)}{(6.7)(0.1)} \text{ m}^3 = 1.11 \text{ m}^3$$

The principle demonstrated in this example is an important one. We may encounter variables that cannot be related to the conversion and we therefore need to relate them to position, in a PFR, or time, in a batch reactor (another demonstration of this can be found in Problem 5.5).

3.3   CSTRs

In an analogous manner to Sections 3.1 and 3.2, we start with the design equation (eqn 2.15) and substitute in for nth-order kinetics,

$$V = \frac{n_{A0} - n_A}{r_A} = \frac{n_{A0}x_A}{kC_A^n} = \frac{n_{A0}x_A v_T^n}{kn_A^n} = \frac{n_{A0}x_A v_T^n}{kn_{A0}^n(1 - x_A)^n}$$

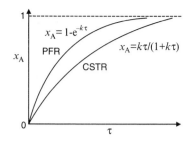

Fig. 3.1 Conversion versus residence time for a PFR and for a CSTR (first-order reaction kinetics).

[5] As we will see in Chapter 6, not all molecules in a CTSR have the same residence time (unlike a PFR or batch reactor) for this reason the volume divided by the volumetric flow rate does not give us an actual residence time but rather a mean residence time.

For a gas-phase reaction we use eqn 3.8,

$$v_T = v_{T0}(1 + \varepsilon_A x_A)$$

$$V = \frac{n_{A0}}{k} \frac{1}{C_{A0}^n} \frac{(1 + \varepsilon_A x_A)^n}{(1 - x_A)^n} x_A$$

If we take a look at the special case of a reaction for which the volumetric flow rate is constant and for which first-order kinetics apply, $\varepsilon_A = 0$, $v_T = v_{T0}$, $n = 1$,

$$\tau = \frac{V}{v_T} = \frac{1}{k} \frac{x_A}{1 - x_A} \tag{3.13}$$

where τ is the mean residence time[5]. Rearranging,

$$x_A = \frac{k\tau}{1 + k\tau} \tag{3.14}$$

As can be seen from the above expression, the conversion in a CSTR approaches the ultimate conversion in a manner that depends upon the reciprocal of mean residence time. This is illustrated in Fig. 3.1. Furthermore, for first-order kinetics a PFR will always need a shorter residence time to reach a given conversion or for a fixed residence time a PFR will always reach a higher conversion.

Expressing eqns 3.13 and 3.14 in terms of the molar flow of reactant,

$$\tau = \frac{1}{k} \frac{n_{A0} - n_A}{n_A} \tag{3.15}$$

$$n_A = \frac{n_{A0}}{1 + k\tau} \tag{3.16}$$

3.4   Comparison of PFRs and CSTRs

We will now make a more detailed comparison of the behaviour of PFRs and CSTRs. Figure 3.2(a) shows a plot of rate versus concentration, i.e. it shows a kinetic dependency, for a reaction with positive-order rate dependence of reactant concentration. Obviously, the outlet concentration of reactant will be lower than the inlet concentration. A CSTR will operate at the outlet condition (as the concentration in the reactor will be the same as the concentration in the exit stream) and the CSTR can therefore be represented by an operating point. Conversely, the PFR needs to be represented by an operating line: the concentration in the reactor continuously changes as we proceed from the inlet to the outlet.

We can now take the same kinetic expression and rearrange it to plot the inverse of the reaction rate versus the conversion (see Fig. 3.2(b)). Again the PFR is represented by an operating line and the CSTR is represented by an operating point. The volume of the PFR is given by,

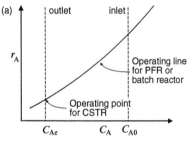

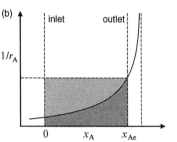

Fig. 3.2 (a) Rate versus concentration for a positive-order reaction. The PFR is represented by an operating line while the CSTR is represented by an operating point. (b) Inverse of reaction rate versus conversion for a positive-order reaction. The shaded area under the curve is proportional to the volume of a PFR while the total shaded area is proportional to the volume of a CSTR.

$$V_{PFR} = n_{A0} \int_0^{x_{Ae}} \frac{dx_A}{r_A}$$

and therefore the volume of the PFR is proportional to the area under the operating line (the heavily shaded area) in Fig. 3.2(b).

The volume of the CSTR is given by,

$$V_{CSTR} = n_{A0}\frac{x_{Ae}}{r_A}$$

and is therefore proportional to the total shaded area (the heavily shaded area plus the lightly shaded area) in Fig. 3.2(b).

From Fig. 3.2(b) we can easily see that, for a given conversion, the volume of a PFR will always be less than that of a CSTR for a positive-order reaction. The PFR tends to operate at higher reactant concentrations that the CSTR, since, in the CSTR, instantaneous dilution with the product takes place. This means that for positive-order reactions, PFRs will exhibit higher overall rates of reaction and therefore will have lower volumes than CSTRs for a given conversion.

Figure 3.3(a) shows a kinetic relationship for a reaction of order less than zero. Figure 3.3(b) shows the same kinetics transposed to show the inverse of the rate versus conversion. It can be seen that for such kinetics the volume of the CSTR (proportional to the lightly shaded area) will be always less than that required with a PFR (the total shaded area). This is because of the 'dilution' of reactants associated with the CSTR which, in this case, results in higher rates.

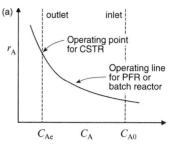

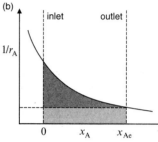

Fig. 3.3 (a) Rate versus concentration for a negative-order reaction. (b) Inverse of reaction rate versus conversion for a negative-order reaction. The total shaded area under the curve is proportional to the volume of a PFR while the lightly shaded area is proportional to the volume of a CSTR.

3.5 Reactors in series

3.5.1 CSTRs of equal volume in series

(a) Calculate the volume of a PFR and a CSTR required for 90% conversion of reactant by a first-order reaction:

$$A \rightarrow B; \quad r_A = kC_A; \quad v_T/k = 1 \text{ m}^3$$

(b) Calculate the total volume of two CSTRs (both of the same volume) in series required for 90% conversion (see Fig. 3.4).

Example 3.4

(a) PFR

$$V = -\int_{n_{A0}}^{n_{Ae}} \frac{dn_A}{r_A} = -\int_{C_{A0}}^{C_{Ae}} \frac{v_T \, dC_A}{k \, C_A} = \frac{v_T}{k} \ln \frac{C_{A0}}{C_{Ae}}$$

$$\frac{C_{A0}}{C_{Ae}} = 10$$

$$V = 2.3 \text{ m}^3$$

CSTR

$$V = \frac{n_{A0} - n_{Ae}}{r_A} = \frac{v_T}{k} \frac{C_{A0} - C_{Ae}}{C_{Ae}} = \frac{v_T}{k}\left(\frac{C_{A0}}{C_{Ae}} - 1\right)$$

$$V = 9 \text{ m}^3$$

(b)

$$V_1 = V_2 = \frac{v_T}{k} \frac{C_{A0} - C_{Ai}}{C_{Ai}} = \frac{v_T}{k} \frac{C_{Ai} - C_{Ae}}{C_{Ae}}$$

$$V_1 = V_2 = \frac{v_T}{k}\left(\frac{C_{A0}}{C_{Ai}} - 1\right) = \frac{v_T}{k}\left(\frac{C_{Ai}}{C_{Ae}} - 1\right)$$

Solution

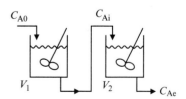

Fig. 3.4 Schematic showing two CSTRs in series.

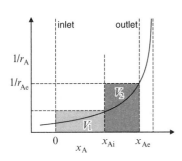

Fig. 3.5 Plot of inverse of reaction rate versus conversion showing that the volume of two CSTRs in series is less that required of one CSTR alone but more than that of a PFR.

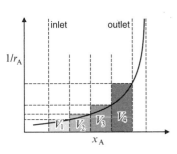

Fig. 3.6 Plot of inverse of reaction rate versus conversion for four CSTRs in series (all with approximately equal conversions).

Therefore,

$$\frac{C_{A0}}{C_{Ai}} = \frac{C_{Ai}}{C_{Ae}}; \qquad \left(\frac{C_{A0}}{C_{Ai}}\right)^2 = \frac{C_{A0}}{C_{Ae}}$$

$$\frac{C_{A0}}{C_{Ai}} = \left(\frac{C_{A0}}{C_{Ae}}\right)^{1/2} = \sqrt{10}$$

$$V_1 = V_2 = (1)(\sqrt{10} - 1) \text{ m}^3 = 2.16 \text{ m}^3$$

Total volume $= V_1 + V_2 + 4.32 \text{ m}^3$

If we look at the two CSTRs of Example 3.4(b) in terms of the plot of inverse rate against conversion, we can see that two CSTRs in series would be expected to have a lower total volume than one CSTR but a greater total volume than that required for a PFR (see Fig. 3.5). Remember that, for the second reactor, the volume depends upon the difference in the outlet conversion and the inlet conversion divided by the rate of reaction in the reactor,

$$V_2 = \frac{n_{A0}(x_{Ae} - x_{Ai})}{r_{Ae}}$$

From Fig. 3.6 (which shows the behaviour of four CSTRs in series, all operating at approximately equal individual conversion), we can see that as the number of CSTRs in series is increased the total volume will eventually approach that of a PFR. This can be shown rigorously in the following way. Consider a first-order irreversible reaction performed with N CSTRs in series (see Fig. 3.7), all of the same volume, V_i, and with a constant volumetric flow rate and the same reaction rate constant, then, from eqn. 3.13,

$$V_i = \frac{v_T}{k} \frac{C_{A0} - C_{A1}}{C_{A1}} = \frac{v_T}{k} \frac{C_{A1} - C_{A2}}{C_{A2}} = \cdots = \frac{v_T}{k} \frac{C_{Ai-1} - C_{Ai}}{C_{Ai}} \cdots = \frac{v_T}{k} \frac{C_{AN-1} - C_{AN}}{C_{AN}}$$

Therefore,

$$\frac{C_{A0}}{C_{A1}} = \frac{C_{A1}}{C_{A2}} = \cdots = \frac{C_{Ai-1}}{C_{Ai}} = \cdots = \frac{C_{AN-1}}{C_{AN}}$$

If we multiply together all of these N relationships we get,

$$\left(\frac{C_{Ai-1}}{C_{Ai}}\right)^N = \frac{C_{A0}}{C_{AN}}$$

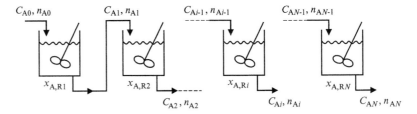

Fig. 3.7 Schematic showing N CSTRs in series.

Substituting into the expression for the volume of any individual reactor,

$$V_i = \frac{v_{\text{T}}}{k}\left[\left(\frac{C_{A0}}{C_{AN}}\right)^{1/N} - 1\right]$$

Hence, the total volume for all N reactors is,

$$V_{\text{total}} = \frac{Nv_{\text{T}}}{k}\left[\left(\frac{C_{A0}}{C_{AN}}\right)^{1/N} - 1\right]$$

As $N \to \infty$

$$V_{\text{total}} = \lim_{N\to\infty} \frac{Nv_{\text{T}}}{k}\left[\left(\frac{C_{A0}}{C_{AN}}\right)^{1/N} - 1\right]$$

But,

$$\left(\frac{C_{A0}}{C_{AN}}\right)^{1/N} = \exp\left\{\frac{1}{N}\ln\left(\frac{C_{A0}}{C_{AN}}\right)\right\}$$

and as $N \to \infty$, because the conversion in the series is finite, then the exponent will tend to zero and therefore the exponential can be re-expressed as a linear approximation,

$$\exp\left\{\frac{1}{N}\ln\left(\frac{C_{A0}}{C_{AN}}\right)\right\} \approx 1 + \frac{1}{N}\ln\left(\frac{C_{A0}}{C_{AN}}\right)$$

Substituting this into the expression for the total volume,

$$V_{\text{total}} = \lim_{N\to\infty} \frac{Nv_{\text{T}}}{k}\left[\frac{1}{N}\ln\left(\frac{C_{A0}}{C_{AN}}\right)\right]$$

$$= \frac{v_{\text{T}}}{k}\ln\left(\frac{C_{A0}}{C_{AN}}\right)$$

This expression is, of course, the same as the expression for a PFR reactor (compare to eqn 3.11). For the same conversion, as the number of CSTRs increases, the total volume approaches that of a PFR (this is true for all types of kinetic behaviour—not just first order). This is consistent with our understanding of these two reactors. Previously we made an analogy between a PFR and infinitesimally small batch reactors. Each differential element in the PFR could be considered a batch reactor as we followed it through the PFR. However, if we fix our frame of reference to a particular place along the length of the PFR and now, instead of following each element, we observe what happens in that fixed element, the element will appear to be an infinitesimally small CSTR—it will be perfectly mixed on this scale and will have flow in and flow out from and to the adjacent CSTRs. Therefore, we would expect to be able to model our PFR as an infinite set of differential CSTR reactors in series.

3.5.2 CSTRs of different volume in series

Now let us consider a number of CSTRs of differing volume that we intend to use in series. How are they best employed to maximize conversion? In the case of first-order kinetics, we have already shown that the conversion in a

CSTR is independent of its feed concentration. For a constant volumetric flow rate the conversion in a CSTR (eqn 3.14) is given by

$$x_A = \frac{k\tau}{1 + k\tau} \quad \text{or,}$$

$$(1 - x_A) = \frac{1}{1 + k\tau}$$

so conversion is dependent only on the rate constant and the mean residence time and not on the inlet concentration. If we have N reactors in series (see Fig. 3.7), we can derive an expression relating the overall conversion to the conversions in the individual reactors.[6]

$$n_{A1} = n_{A0}(1 - x_{A,R1})$$
$$n_{A2} = n_{A1}(1 - x_{A,R2}) \quad \text{and}$$
$$n_{AN} = n_{AN-1}(1 - x_{A,RN})$$

where $x_{A,Ri}$ refers to the conversion of reactant A in the ith reactor.
    Therefore,

$$n_{AN} = n_{A0}(1 - x_{A,R1})(1 - x_{A,R2})\ldots(1 - x_{A,RN})$$

and the overall conversion of the series, x_A, is given by,

$$x_A = 1 - \frac{n_{AN}}{n_{A0}} = 1 - (1 - x_{A,R1})(1 - x_{A,R2})\ldots(1 - x_{A,RN}) \quad \text{or,}$$
$$(1 - x_A) = (1 - x_{A,R1})(1 - x_{A,R2})\ldots(1 - x_{A,RN}) \quad (3.17)$$

Therefore, for N CSTRs in series with a first-order reaction,

$$(1 - x_A) = \frac{1}{(1 + k\tau_1)(1 + k\tau_2)\ldots(1 + k\tau_N)}$$

and we can see that the order of the individual reactors cannot influence the overall conversion.

    What happens with a first-order reaction is a limiting case. For other reaction orders the sequence of reactors is important. If the reaction kinetics are of order greater than one it is now important for us to keep concentrations as high as possible for as long as possible to benefit from the associated high rates of reaction. Therefore, we should place the smallest CSTRs first (see Fig. 3.8(a)). Conversely, if the reaction is of order less than one it is advantageous to lower reactant concentrations as early as possible. Therefore, the largest CSTRs should be placed first (see Fig. 3.8(b)).

    There is also the special case where the reaction is zero order and the size of individual reactors becomes unimportant (and therefore the order must be unimportant); the only thing that matters is the total reactor volume available[7].

3.5.3   CSTRs of optimum volume in series

Now let us consider how to minimize the total reactor volume (for a given conversion) if we are allowed to choose the sizes of individual reactors in the reactor chain (a single optimum exists in all cases for $n > 0$; for the zero-order case only the total volume is important[8]). For instance, consider using two CSTRs to achieve a particular conversion in a system displaying positive-

[6] The analysis does not only apply to CSTRs as shown.

[7] We will not discuss the case of negative reaction orders. Multiple solutions are available and analysis is more complex.

[8] Again the analysis for negative reaction orders is beyond the scope of this text.

(a)

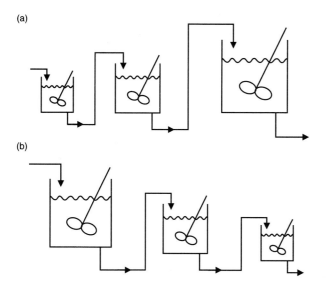

(b)

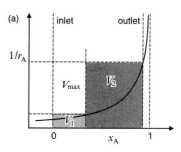

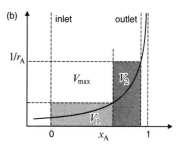

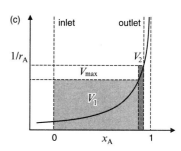

Fig. 3.8   (a) For reactions with positive-order kinetics small CSTRs should be used early to maximize conversion. (b) For reactions with negative-order kinetics large CSTRs should be used early to maximize conversion.

Fig. 3.9   Plot of inverse of reaction rate versus conversion showing the volume of two CSTRs in series; (a) the first reactor is much smaller than the second; (b) the reactors are of similar size;   (c) the second reactor is much smaller than the first.

order kinetics. Figure 3.9 shows three choices of reactor sizes. In all three cases,

$$V_{max} + V_1 + V_2 = \text{constant}$$

where the constant is the volume that would be required for one CSTR operating alone. V_1 and V_2 are the volumes of the individual reactors and, therefore, if we are trying to minimize their sum we should maximize V_{max}. In Fig. 3.9(a) the first reactor is much smaller than the second, in Fig. 3.9(b) the reactors are of a similar size, and in Fig. 3.9(c) the second reactor is much smaller than the first. It can clearly be seen that the intermediate case gives the smallest total reactor volume. In fact, it can be shown (see Example 3.5) that for first-order kinetics the reactors should be chosen to be of equal size to minimize the total volume. For kinetics of order $n > 1$, the first reactor should be smaller than the second reactor. For $n < 1$, the first reactor should be larger than the second. The saving in total reactor volume by having reactors of different size tends to be very modest and is therefore, in general, not worthwhile economically.

A given conversion is to be achieved using two CSTRs in series. The total reactor volume is to be minimized. Show that, for first-order kinetics, the reactors should be chosen to be of equal size.

Example 3.5

For a CSTR (eqn 3.14),

Solution

$$x_A = \frac{k\tau}{1 + k\tau}$$

$$(1 - x_A) = \frac{1}{1 + k\tau}$$

For two reactors in series (eqn 3.17),

$$(1 - x_A) = (1 - x_{A,R1})(1 - x_{A,R2})$$

For two CSTRs,

$$(1 - x_A) = \frac{1}{(1 + k\tau_1)(1 + k\tau_2)}$$

$$\frac{1}{(1 - x_A)} = (1 + k\tau_1)(1 + k\tau_2)$$

Hence, for a given conversion, x_A, since k is a constant, τ_1 is a function of τ_2 alone. We wish to minimize the total volume or total residence time, $(\tau_1 + \tau_2)$, by choosing an appropriate value of τ_2,

$$\frac{d}{d\tau_2}(\tau_1 + \tau_2) = 0$$

$$\frac{d\tau_1}{d\tau_2} + 1 = 0; \quad \frac{d\tau_1}{d\tau_2} = -1$$

Remembering that,

$$\frac{1}{(1 - x_A)} = (1 + k\tau_1)(1 + k\tau_2)$$

and differentiating with respect to τ_2, with x_A constant,

$$0 = k\frac{d\tau_1}{d\tau_2}(1 + k\tau_2) + k(1 + k\tau_1)$$

Substituting for $\dfrac{d\tau_1}{d\tau_2}$

$$k(1 + k\tau_2) = k(1 + k\tau_1) \quad \text{or,}$$

$$\tau_2 = \tau_1$$

3.5.4   PFR and CSTR in series

When PFRs and CSTRs are both available and to be used in series, the order of the reactors can affect the total conversion. The conversion in any one reactor will, in general, be dependent upon the inlet concentration to that reactor. Therefore, the overall conversion for a CSTR and PFR in series will, in general, depend upon the order in which the reactors are placed since the order will affect the inlet concentration to each reactor. There are two exceptions to this. The overall conversion for a CSTR and a PFR in series is independent of reactor sequence for first-order reactions as the individual reactor conversions are independent of inlet concentration. The overall conversion is also independent of sequence for zero-order reactions as only the total reactor volume available is important.

Let us consider the case of first-order reactions. We have already seen that for a constant volumetric flow rate the conversion in a CSTR (eqn 3.14) is given by,

$$x_{A,CSTR} = \frac{k\tau_{CSTR}}{1 + k\tau_{CSTR}}$$

which is dependent only on the rate constant and the mean residence time. The conversion in a PFR (eqn 3.10) is given by,

$$x_{A,PFR} = 1 - e^{-k\tau_{PFR}}$$

The total conversion for the combined CSTR and PFR is then given by (using eqn 3.17),

$$(1 - x_A) = (1 - x_{A,CSTR})(1 - x_{A,PFR}) = \frac{e^{-k\tau_{PFR}}}{1 + k\tau_{CSTR}}$$

As the conversion in both a PFR and a CSTR is independent of inlet concentration, the total conversion for the two reactors in series must be the same regardless of the order in which the reactors are placed.

For reaction orders greater than one, high rates (at high concentrations) should be taken advantage of if conversion is to be maximized. As a PFR takes advantage of the high reaction rates it should be placed before the CSTR. If more than one CSTR is available, then the smallest ones should be used first (immediately after the PFR) so as to keep concentrations higher. However, if the reaction order is less than one, then it is better to dilute the stream as early as possible. Therefore, the conversion would be greater if the CSTR were placed before the PFR.

A CSTR and a PFR are to be used in series to decompose a reactant A via a second-order process with rate constant of 0.002 litre mol^{-1} s^{-1}. The reaction is performed in the liquid phase. The inlet concentration of A to the reactor is 5 mol litre^{-1} at a flow rate of 0.02 litre s^{-1}. The volume of the CSTR is 2 litres and the volume of the PFR is 2 litres. What order should the CSTR and PFR be placed in to maximize conversion? What is this maximum conversion? **Example 3.6**

The PFR should be placed first to take advantage of the higher concentrations and therefore reaction rates. **Solution**

$$r_A = -\frac{dn_A}{dV} = -v_T\frac{dC_A}{dV} = kC_A^2$$

$$V = -v_T\int_{C_{A0}}^{C_{Ai}}\frac{dC_A}{kC_A^2} = \frac{v_T}{k}\left[\frac{1}{C_A}\right]_{C_{A0}}^{C_{Ai}}$$

where C_{Ai} is the concentration between the reactors,

$$V = \frac{v_T}{k}\frac{C_{A0} - C_{Ai}}{C_{A0}C_{Ai}}$$

$$\frac{C_{Ai}}{C_{A0}} = \frac{1}{1 + k\tau C_{A0}}$$

As $k\tau C_{A0} = 1$, $C_{Ai} = 0.5C_{A0}$

The PFR is then followed by a CSTR,

$$V = \frac{v_T(C_{Ai} - C_{Ae})}{kC_{Ae}^2}$$

$$k\tau C_{Ae}^2 = C_{Ai} - C_{Ae}$$

$$k\tau C_{Ae}^2 + C_{Ae} - C_{Ai} = 0$$

$$C_{Ae} = \frac{-1 \pm \sqrt{1 + 4k\tau C_{Ai}}}{2k\tau}$$

$$C_{Ae} = \left(\frac{-1 \pm \sqrt{1 + 2}}{2}\right) C_{A0}$$

$$C_{Ae} = 0.366 C_{A0}$$

63.4% conversion

If the CSTR were placed before the PFR, then the conversion achieved would be less.

$$V = \frac{v_T(C_{A0} - C_{Ai})}{k C_{Ai}^2}$$

$$C_{Ai} = \frac{-1 \pm \sqrt{1 + 4k\tau C_{A0}}}{2k\tau}$$

$$C_{Ai} = \left(\frac{-1 \pm \sqrt{5}}{2}\right) C_{A0} = 0.618 C_{A0}$$

This is followed by the PFR,

$$r_A = -\frac{dn_A}{dV} = -v_T \frac{dC_A}{dV} = k C_A^2$$

$$V = -v_T \int_{C_{Ai}}^{C_{Ae}} \frac{dC_A}{k C_A^2} = \frac{v_T}{k}\left[\frac{1}{C_A}\right]_{C_{Ai}}^{C_{Ae}}$$

$$V = \frac{v_T}{k} \frac{C_{Ai} - C_{Ae}}{C_{Ai} C_{Ae}}$$

$$\frac{C_{Ae}}{C_{Ai}} = \frac{1}{1 + k\tau C_{Ai}} = \frac{1}{1 + 0.618 k\tau C_{A0}} = \frac{1}{1 + 0.618} = 0.618$$

$$C_{Ae} = (0.618)(0.618) C_{A0} = 0.382 C_{A0}$$

This gives an overall conversion of 61.8%.

We can see that there is an advantage to be gained if the PFR is placed before the CSTR for a second-order reaction. However, the increase in conversion is modest (63.4% versus 61.8%).

3.6   The recycle reactor

Figure 3.10 depicts a recycle reactor. Fresh feed is mixed with a recycle stream and then fed to the PFR. The outlet stream from the reactor is split; part of the stream is recycled to the reactor inlet and the rest is the product stream. We define the recycle ratio in the following way,

$$R = \frac{\text{moles returning}}{\text{moles leaving system}}$$

If we first look at the reactor alone and ignore the fact that there is a recycle, we can write down an expression for conversion (known as the per-pass conversion),

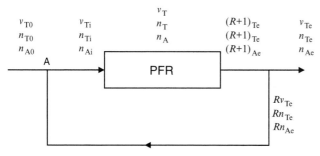

Fig. 3.10 Schematic of a recycle reactor.

$$x_A = \frac{n_{Ai} - (R+1)n_{Ae}}{n_{Ai}} \tag{3.18}$$

If we now treat the reactor and its recycle to be one 'overall' reactor then, looking at the inlet and outlet streams, we can write down an expression for the overall conversion,

$$X_A = \frac{n_{A0} - n_{Ae}}{n_{A0}}; \quad n_{Ae} = n_{A0}(1 - X_A) \tag{3.19}$$

We now need to undertake some process analysis to understand the behaviour of the system. Performing a material balance at point A in Fig. 3.10,

$$n_{Ai} = n_{A0} + Rn_{Ae} \tag{3.20}$$

Substituting eqn 3.19 into eqn 3.20,

$$n_{Ai} = n_{A0}(1 + R - RX_A) \tag{3.21}$$

Substituting eqn 3.21 into eqn 3.18,

$$x_A = \frac{n_{A0}(1 + R - RX_A) - n_{A0}(1 + R)(1 - X_A)}{n_{A0}(1 + R - RX_A)}$$

$$x_A = \frac{X_A}{(1 + R - RX_A)} \tag{3.22}$$

So when $R = 0$, $x_A = X_A$, i.e. per-pass conversion and overall conversion are the same thing (which is of course to be expected as the reactor is then simply an ordinary PFR). As R tends to infinity for finite X_A then x_A must tend to zero[9].

Now we need to be able to calculate the reactor volume. The volume will be given by the usual integral from the inlet condition to outlet condition,

$$V = -\int_{n_{Ai}}^{(R+1)n_{Ae}} \frac{dn_A}{r_A} \tag{3.23}$$

In performing the integration we will leave the variable as n_A because the use of conversion as the variable can introduce confusion (the inlet conversion to the reactor is not zero as part of the product stream is being recycled).

For a first-order reaction,

$$r_A = kC_A = k\frac{n_A}{v_T}$$

[9] The fraction recycled, α, is also sometimes used,

$$\alpha = \frac{R}{R+1}$$

Rearranging,

$$R = \frac{\alpha}{1 - \alpha}$$

Substituting into eqn 3.22.

$$x_A = \frac{X_A}{\left(1 + \frac{\alpha}{1-\alpha} - \frac{\alpha}{1-\alpha}X_A\right)}$$
$$= \frac{(1 - \alpha)X_A}{(1 - \alpha X_A)}$$

Let us assume that we are dealing with a reaction for which there is no volume change, i.e. a liquid-phase reaction or a gas-phase reaction with an equal number of moles on both sides of the reaction. Then,

$$v_{T0} = v_{Te}$$

$$v_{Ti} = v_T = (R+1)v_{Te} = (R+1)v_{T0}$$

$$r_A = kC_A = k\frac{n_A}{v_T} = k\frac{n_A}{(R+1)v_{T0}}$$

Substituting the above into eqn 3.23, the integral for volume becomes,

$$V = -\int_{n_{Ai}=n_{A0}+Rn_{Ae}}^{(R+1)n_{Ae}} \frac{(R+1)v_{T0}}{k}\frac{dn_A}{n_A}$$

$$= -\frac{(R+1)v_{T0}}{k}\left[\ln n_A\right]_{n_{A0}+Rn_{Ae}}^{(R+1)n_{Ae}}$$

$$= \frac{(R+1)v_{T0}}{k}\ln\frac{n_{A0}+Rn_{Ae}}{(R+1)n_{Ae}}$$

Now we will consider what happens in the limits of R tending to zero and infinity.

As R tends to zero,

$$V = \frac{v_{T0}}{k}\ln\frac{n_{A0}}{n_{Ae}}$$

as would be expected because the reactor behaves as a simple PFR (compare to eqn 3.11).

As R tends to infinity,

$$V = \frac{(R+1)v_{T0}}{k}\ln\left\{1+\frac{n_{A0}-n_{Ae}}{(R+1)n_{Ae}}\right\}$$

As R increases this exponential can be approximated; for small x, $\ln(1+x) \approx x$,

$$V = \frac{(R+1)v_{T0}}{k}\frac{n_{A0}-n_{Ae}}{(R+1)n_{Ae}} = \frac{v_{T0}}{k}\frac{n_{A0}-n_{Ae}}{n_{Ae}}$$

This is of course the expression that would be expected for a CSTR (compare to eqn 3.15). So a PFR with an infinite recycle ratio behaves as a CSTR. By using the recycle we are effectively inducing mixing in the reactor. As the recycle ratio approaches infinity the concentration gradients within the reactor become negligible (as far as determining reaction rates are concerned) and this is analogous to perfect mixing. Therefore, we would expect the recycle reactor to behave like a CSTR. This is consistent with eqn 3.22 which shows that as R is increased the per-pass conversion must go to zero, i.e. concentration gradients disappear.

Example 3.7   For a first-order, irreversible reaction,

$$A \rightarrow B; r_A = kC_A; k\tau = 2$$

sketch x_A (per-pass conversion) and X_A (total conversion) versus fraction recycled, α, for a PFR with recycle.

PFR: recall eqn 3.10,

$$x_A = 1 - e^{-k\tau}$$

CSTR: recall eqn 3.14

$$x_A = \frac{k\tau}{1 + k\tau}$$

$\alpha = 0$, overall conversion is the same as per-pass conversion, $x_A = X_A = 1 - e^{-2}$
$\alpha = 1$, there is negligible per-pass conversion and the reactor gives an overall conversion the same as that for a CSTR, $x_A = 0; X_A = 2/3$
See Fig. 3.11.

3.7 Problems

3.1 A gas-phase reaction is performed in a batch reactor (constant volume),

$$A \rightarrow 2B \quad r_A = kP_A$$

Initially, 50% A and 50% inert are present on a volume basis.

$P_0 = 1.00$ bar, $P_\theta = 1.38$ bar,

$t = 170$ s, $T = 500$ K,

$R = 8.314$ J mol^{-1} K^{-1}

(a) What is the final conversion?
(b) Show that $r_A V = kRTN_A$. (c) What is the value of k?

3.2 It is intended to carry out a gas-phase dimerization reaction in an isothermal batch reactor,

$$2A \rightarrow B$$

The reaction is second order: $r_A = kC_A^2$
A conversion of 90% must be achieved.

Calculate the ratio of the residence time of a constant volume reactor to the residence time of a constant pressure reactor. Initially, both reactors are of the same volume and at the same pressure, and no product, B, is present.

Why is the ratio of residence times greater than, or less than, unity? (Both reactors operate at the same temperature and the mixture behaves as a perfect gas.)

3.3 A gas-phase reaction is performed in a CSTR,

$$2A \Leftrightarrow B \quad r_A = k_1 P_A^2 - k_{-1}P_B$$

$$\frac{n_{A0}}{n_{T0}} = \frac{2}{3}; \frac{n_{B0}}{n_{T0}} = \frac{1}{3}$$

$P = 1$ bar; $k_1 = 0.05$ mol m^{-3} s^{-1} bar^{-2}, $k_{-1} = 0.025$ mol m^{-3}s^{-1} bar^{-1}, v_{T0} (600 K, 1 bar) $= 1$ litre s^{-1}, $T = 600$ K, $R = 8.314$ J mol^{-1} K^{-1}

(a) What are $P_A,$ P_B at $x_A = 0.25$?
(b) What is the reaction rate at $x_A = 0.25$? (c) How many moles of A are fed to the reactor, i.e. n_{A0}? (d) What reactor volume is required for $x_A = 0.25$?

3.4 A second-order liquid-phase reaction is to be carried out in a PFR:

$$A + B \rightarrow \text{Products} \quad r_A = kC_A C_B;$$

$$k = 10^{-6} \text{ m}^3 \text{ s}^{-1} \text{ mol}^{-1}$$

The inlet concentration of both A and B is 10^3 mol m^{-3} and the flow rate is 10^{-5} m^3 s^{-1}. Calculate the reactor volume required for 50% conversion.

3.5 Consider a gas-phase reaction performed in an isothermal PFR,

$$A + B \rightarrow C$$

which is zero order with respect to A and first order with respect to B:

$$r_A = kP_B; \quad k = 2.3 \times 10^3 \text{ mol s}^{-1} \\ \text{m}^{-3} \text{ bar}^{-1}$$

At a total feed rate of 120 mol s^{-1} calculate the reactor volume required to obtain 45% conversion of A. The feed is A and B in a molar ratio of 2:1. The total operating pressure is 20 bar.

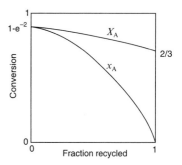

Fig. 3.11 Plot of conversion versus fraction recycled for a PFR with recycle. Both per-pass conversion, x_A and overall conversion, X_A, are shown.

3.6   The first-order, reversible reaction,

$$A \Leftrightarrow B \quad r_A = k_1 P_A - k_{-1} P_B$$

is conducted in an isothermal PFR of volume $2 \times 10^{-3} m^3$. The total pressure is 5 bar. The rate constants are 1.2 and 0.8 mol m^{-3} s^{-1} bar for the forward and reverse reactions respectively. The feed is pure A at a rate of 0.01 mol s^{-1}.

(a) Calculate the percentage conversion of A. (b) What will be the new overall conversion if half of the reactor outlet is recycled and mixed with the feed? (c) What is the limiting conversion reached as the fraction recycled is increased?

4 Multiple reactions

Up until now we have only considered what happens if just one reaction is taking place in the reactor. However, the desired reaction is often one of a number of reactions occurring. Consider for example the oxidation of ethylene in air. The desired product is ethylene oxide (Reaction R4.1) but total oxidation products, carbon dioxide and water, are produced through Reaction R4.2. Furthermore, our desired product can itself further react to total oxidation products through Reaction R4.3.

$$C_2H_4 + \frac{1}{2}O_2 \rightarrow C_2H_4O \tag{R4.1}$$

$$C_2H_4 + 3O_2 \rightarrow 2CO_2 + 2H_2O \tag{R4.2}$$

$$C_2H_4O + 2\frac{1}{2}O_2 \rightarrow 2CO_2 + 2H_2O \tag{R4.3}$$

In a case such as this, the important performance indicator is the amount of ethylene oxide produced and not the amount of ethylene that is consumed. Reactions R4.1 and R4.2 are said to be in parallel; they both compete for the same reactant. Reactions R4.1 and R4.3 are said to be in series; ethylene oxide must be produced by Reaction R4.1 before it can be consumed by Reaction R4.3.

To proceed it is helpful to introduce some definitions. Consider a general reaction network which includes both series and parallel reactions,

$$\nu_{A1}A \rightarrow \nu_{B1}B + \nu_{C1}C$$

$$\nu_{A2}A \rightarrow \nu_{D2}D$$

$$\nu_{B3}B \rightarrow \nu_{E3}E$$

where ν_{A1} is the stoichiometric coefficient of A in reaction 1, etc. Using A as the reference reactant we can define the yield and selectivity of the products (here we will use the notation corresponding to a continuous reactor). The yield of B (with reference to the amount of reactant, A, fed to the reactor), $Y_{B/A}$, is given by,

$$Y_{B/A} = \frac{n_B - n_{B0}}{\nu_{B1}} \frac{\nu_{A1}}{n_{A0}} \tag{4.1}$$

i.e. the yield of B is the number of moles of B formed for every mole of A fed to the reactor, with the stoichiometric coefficients chosen so that if there are no competing reactions the yield will be equal to the conversion of A[1].

The overall selectivity for B (with reference to the amount of reactant, A, that has disappeared), $S_{B/A}$, is given by,

$$S_{B/A} = \frac{n_B - n_{B0}}{\nu_{B1}} \frac{\nu_{A1}}{n_{A0} - n_A} \tag{4.2}$$

[1] This is the same as saying that the yield of B is the number of moles of A required (to produce the B), $\nu_{A1}(n_B - n_{B0})/\nu_{B1}$, divided by the total amount of A fed to the reactor, n_{A0}. If we compare this definition of yield to the definition of conversion (the total number of moles of A reacted divided by the total A fed to the reactor), we can see that yield and conversion are closely related (as previously mentioned they are the same if there are no competing reactions). In fact, the yield can be viewed as an individual conversion to a given product.

i.e. the overall selectivity for B is the number of moles of B formed for every mole of A reacted, with the stoichiometric coefficients chosen so that if there are no competing reactions the overall selectivity will be 100%[2].

Using eqns 4.1 and 4.2 the yield and overall selectivity can be related,

$$Y_{B/A} = x_A S_{B/A} \tag{4.3}$$

Rather than evaluate selectivity based on the reactant and product flows in and out of the reactor, we can also evaluate selectivity at any point in the reactor (continuous reactor) or at any time (batch reactor). This selectivity is known as the local (continuous) or instantaneous (batch) selectivity, $\phi_{B/A}$,

$$\phi_{B/A} = -\frac{\nu_{A1}}{\nu_{B1}} \frac{dn_B}{dn_A} \tag{4.4}$$

Note that eqn 4.4 could be arrived at by writing eqn 4.2 for a differential element of reactor length (PFR) or differential element of time (batch reactor).

The yield, overall selectivity, and local (or instantaneous) selectivity are defined in the same manner for all of the other products[3].

4.1 Parallel reactions

Now consider the simple reaction scheme,

$$A \rightarrow R \qquad r_{A1} = k_1 C_A \tag{R4.4}$$

$$A \rightarrow 2S \qquad r_{A2} = k_2 C_A \tag{R4.5}$$

where r_{A1} is the rate of disappearance of A through Reaction R4.4 or reaction 1, and r_{A2} is the rate of disappearance of A through Reaction R4.5 or reaction 2. Therefore, the total rate of disappearance of A, r_A, is simply the sum of the two individual rates,

$$r_A = r_{A1} + r_{A2} \tag{4.5}$$

For simple parallel reactions (i.e. only one product produced per reaction), the sum of the individual product yields must equal the conversion of reactant,

$$x_A = \sum_{i=\text{products}} Y_{i/A} \tag{4.6}$$

Equation 4.6 is true because the yield of each product is simply equal to the fraction of reactant that reacts to produce that particular product. So if we sum all of these fractions we must recover the total fraction of reactant that disappears.

The proportion of the total reaction which produces the desired product is the overall selectivity for that product,

$$S_{R/A} = \frac{Y_{R/A}}{x_A} = \frac{Y_{R/A}}{Y_{R/A} + Y_{S/A}} \tag{4.7}$$

$$S_{S/A} = \frac{Y_{S/A}}{x_A} = \frac{Y_{S/A}}{Y_{R/A} + Y_{S/A}} \tag{4.8}$$

---

[2] This is the same as saying that the selectivity for B is the number of moles of A required (to produce the B), $\nu_{A1}(n_B - n_{B0})/\nu_{B1}$, divided by the total amount of A reacted, $n_{A0} - n_A$.
[3] It is best to always think of yield and selectivity in terms of the amount of reactant required to make the product (rather than the amount of product formed) divided by the reactant supplied (yield) or the reactant consumed (selectivity). This avoids problems associated with the stoichiometry.

The instantaneous or local selectivity is based upon the instantaneous or local reaction rates,

$$\phi_{R/A} = \frac{r_{A1}}{r_A} = \frac{r_{A1}}{r_{A1} + r_{A2}} \tag{4.9}$$

$$\phi_{S/A} = \frac{r_{A2}}{r_A} = \frac{r_{A2}}{r_{A1} + r_{A2}} \tag{4.10}$$

4.1.1 Batch reactors and PFRs

Consider again Reactions R4.4 and R4.5. For a batch reactor (PFR behaviour is analogous except that residence time is used instead of reaction time),

$$r_A = -\frac{1}{V}\frac{dN_A}{dt} = r_{A1} + r_{A2} = k_1 C_A + k_2 C_A \tag{4.11}$$

As
$$C_A = \frac{N_A}{V}$$

$$\int_{N_{A0}}^{N_A} \frac{dN_A}{N_A} = -\int_0^t (k_1 + k_2)dt$$

$$N_A = N_{A0} \exp\{-(k_1 + k_2)t\} \tag{4.12}$$

And we can see that A will disappear with a time constant that depends upon the inverse of the sum of the rate constants (compare to eqn 3.5 for a single reaction). What we are more interested in is how the number of moles of R and S will vary with time.

From eqns 4.9 and 4.10 the instantaneous selectivities are given by,

$$\phi_{R/A} = \frac{r_{A1}}{r_{A1} + r_{A2}} = \frac{k_1}{k_1 + k_2} \quad \text{and;} \tag{4.13}$$

$$\phi_{S/A} = \frac{r_{A2}}{r_{A1} + r_{A2}} = \frac{k_2}{k_1 + k_2} \tag{4.14}$$

The instantaneous selectivities tell us what fraction of reactant is consumed by each reaction, i.e. what fraction of reactant is responsible for formation of each product at any given time. Note that, in this case, both of the instantaneous selectivities are constant and, therefore, the overall selectivities must be equal to the local selectivities. Let us show this rigorously by considering the whole reactor. We can form two design equations, one each for the two individual reactions,

$$r_{A1} = \frac{1}{V}\frac{dN_{A1}}{dt} = k_1 C_A \tag{4.15}$$

$$r_{A2} = \frac{1}{V}\frac{dN_{A2}}{dt} = k_2 C_A \tag{4.16}$$

where N_{A1} and N_{A2} are the number of moles of A that disappear through reactions 1 and 2 respectively[4]. If we add eqns 4.15 and 4.16 together,

$$r_{A1} + r_{A2} = \frac{1}{V}\frac{dN_{A1}}{dt} + \frac{1}{V}\frac{dN_{A2}}{dt} = k_1 C_A + k_2 C_A \tag{4.17}$$

---

[4] N_{A1} and N_{A2} are closely related, through the stoichiometry, to the amounts of products formed and equations 4.15 and 4.16 are therefore similar to design equations for products.

But because, $$N_A = N_{A0} - N_{A1} - N_{A2} \qquad (4.18)$$

$$\frac{dN_A}{dt} = -\frac{dN_{A1}}{dt} - \frac{dN_{A2}}{dt}$$

If the above equation is substituted into eqn 4.17, then eqn 4.11 is recovered. However, by treating the two reactions separately we can get the extra information that is required to know the product distribution.

We can divide the design equations, eqns 4.15 and 4.16, by one another,

$$\frac{dN_{A1}}{dN_{A2}} = \frac{k_1}{k_2} \qquad (4.19)$$

Integrating eqn 4.19 with the initial condition, $N_{A1} = N_{A2} = 0$,

$$\frac{N_{A1}}{N_{A2}} = \frac{k_1}{k_2} \qquad (4.20)$$

The above equation applies for any set of irreversible parallel reactions with equal kinetic orders. In such a case, the moles of reactant that disappear due to a particular reaction are simply proportional to the rate constant. This means that if we want to change the reaction selectivity we can only do this by changing temperature and so changing rate constants[5].

We can now proceed to obtain expressions for the number of moles of A that disappear through each individual reaction,

As $N_A = N_{A0} - N_{A1} - N_{A2}$

using eqn 4.20,

$$N_A = N_{A0} - N_{A1}\left(1 + \frac{k_2}{k_1}\right)$$

$$N_{A1} = \frac{k_1}{k_1 + k_2}(N_{A0} - N_A) \qquad (4.21)$$

$$N_{A2} = \frac{k_2}{k_1 + k_2}(N_{A0} - N_A) \qquad (4.22)$$

The overall selectivity for R is given by the number of moles of A that disappear through reaction 1 divided by the total number of moles of A that disappear through both reactions,

$$S_{R/A} = \frac{N_{A1}}{N_{A1} + N_{A2}} = \frac{N_{A1}}{N_{A0} - N_A} = \frac{k_1}{k_1 + k_2} \qquad (4.23)$$

and the overall selectivity for S is given by,

$$S_{S/A} = \frac{N_{A2}}{N_{A1} + N_{A2}} = \frac{N_{A2}}{N_{A0} - N_A} = \frac{k_2}{k_1 + k_2} \qquad (4.24)$$

The overall selectivity is therefore equal to the instantaneous selectivity (compare eqns 4.23 and 4.24 to eqns 4.13 and 4.14) as expected. Although this may seem trivial, the principle demonstrated is important. The instantaneous or local selectivity depends upon the instantaneous or local

---

[5] If all of the reactions occurring are of equal reaction rate orders, then in a batch reactor,

$$\frac{N_{A1}}{N_{A2}} = \frac{k_1}{k_2}$$

and in a continuous reactor,

$$\frac{n_{A1}}{n_{A2}} = \frac{k_1}{k_2}$$

and the selectivities can only be modified by changing the rate constants (the choice of reactor cannot influence selectivity). Rate constants are a function of temperature alone,

$$k_1 = k'_1 \exp\left(-\frac{\Delta E_1}{RT}\right)$$

$$k_2 = k'_2 \exp\left(-\frac{\Delta E_2}{RT}\right)$$

where k'_1 and k'_2 represent the appropriate Arrhenius pre-exponential factors and ΔE_1 and ΔE_2 are the activation energies associated with reactions 1 and 2 respectively.

Therefore, the rate of one reaction relative to another can be modified by changing the temperature,

$$\frac{N_{A1}}{N_{A2}} = \frac{k_1}{k_2} = \frac{k'_1}{k'_2} \exp\left\{\frac{(\Delta E_2 - \Delta E_1)}{RT}\right\}$$

If $\Delta E_1 > \Delta E_2$ then an increase in temperature will increase reaction rate 1 more than reaction rate 2 and the selectivity for the first reaction will increase.

reaction rates while the overall selectivity depends upon integration of the design equation for the relevant reaction.

It is now straightforward to get expressions for the moles of R and S present[6] as a function of N_{A1} and N_{A2}.

$$N_R = N_{R0} + N_{A1}$$

Using eqn 4.21,

$$N_R = N_{R0} + \frac{k_1}{k_1 + k_2}(N_{A0} - N_A)$$

and we already know how the amount of A depends upon time (eqn 4.12), so substituting,

$$N_R = N_{R0} + \frac{k_1}{k_1 + k_2}N_{A0}\left[1 - \exp\{-(k_1 + k_2)t\}\right]$$

Likewise, $N_S = N_{S0} + 2N_{A2}$

$$= N_{S0} + \frac{2k_2}{k_1 + k_2}N_{A0}\left[1 - \exp\{-(k_1 + k_2)t\}\right]$$

Fig. 4.1 Reaction network from Example 4.1.

Consider the reaction scheme in Fig. 4.1. All reactions are first order in A with rate constants,

$$k_1 = 1 \times 10^{-3}\,s^{-1},\ k_2 = 2 \times 10^{-3}\,s^{-1},\ k_3 = 3 \times 10^{-3}\,s^{-1}$$

What residence time is required to achieve a 10% yield of B in a batch reactor?

Example 4.1

Orders are equal so, $\begin{cases} \dfrac{k_1}{k_2} = \dfrac{1}{2} = \dfrac{N_{A1}}{N_{A2}} = \dfrac{N_B}{N_C};\ \dfrac{k_1}{k_3} = \dfrac{1}{3} = \dfrac{N_{A1}}{N_{A3}} = \dfrac{N_B}{N_D} \end{cases}$

Solution

$$N_C = 2N_B;\quad N_D = 3N_B$$

Because the rate of reaction 2 is twice the rate of reaction 1, two moles of C will be produced for each mole of B produced. Furthermore, because of the one-to-one stoichiometry, the number of moles of A consumed must equal the total number of moles of B, C, and D produced.

$$N_{A0}x_A = N_B + N_C + N_D = 6N_B$$

We require a 10% yield of B,

$$\frac{N_{A1}}{N_{A0}} = \frac{N_B}{N_{A0}} = Y_{B/A} = 0.1$$

$$x_A = \frac{N_B + N_C + N_D}{N_{A0}} = 6\frac{N_B}{N_{A0}} = 0.6$$

$$\left(Y_{B/A} = 0.1;\ Y_{C/A} = 0.2;\ Y_{D/A} = 0.3;\ S_{B/A} = \frac{1}{6}\right)$$

$$N_A = N_{A0}\exp[-(k_1 + k_2 + k_3)t]$$

[6] Notice that we have not had to use the reaction stoichiometries for any of this analysis until we finally calculate the amounts of the products formed.

$$t = -\frac{1}{k_1 + k_2 + k_3} \ln \frac{N_A}{N_{A0}} = -\frac{1}{6 \times 10^{-3}} \ln(0.4) = 153 \text{ s}$$

Example 4.2 The following reactions are performed isothermally in the gas phase in a PFR,

$$A \rightarrow B \qquad r_{A1} = k_1 P_A$$

$$A \rightarrow 2C \qquad r_{A2} = k_2 P_A$$

$$A \rightarrow 3D \qquad r_{A3} = k_3 P_A$$

The reaction rate constants are given by,

$$k_1 = 5.4 \times 10^2 \exp\left(\frac{-20\,000}{RT}\right) \text{mol s}^{-1} \text{m}^{-3}\text{bar}^{-1}$$

$$k_2 = 6.5 \times 10^3 \exp\left(\frac{-40\,000}{RT}\right) \text{mol s}^{-1} \text{m}^{-3}\text{bar}^{-1}$$

$$k_3 = 2.1 \times 10^2 \exp\left(\frac{-20\,000}{RT}\right) \text{mol s}^{-1} \text{m}^{-3}\text{bar}^{-1}$$

where RT is in J mol^{-1}.

(a) At what temperature should the reactor be operated to achieve a 20% selectivity for D (i.e. one mole of A reacts through reaction 3 for every five moles of A reacted)?

(b) At this temperature, what reactor volume is required for a 10% yield of D (i.e. one mole of A reacts through reaction 3 for every ten moles of A fed to the reactor) if pure A is fed to the reactor at a rate of 2 mol s^{-1}. The reactor pressure is 1 bar.

Solution (a) $\phi_{D/A} = \dfrac{r_{A3}}{r_A + r_{A2} + r_{A3}} = \dfrac{k_3}{k_1 + k_2 + k_3} = S_{D/A} = 0.2$

$$0.8k_3 - 0.2k_1 = 0.2k_2$$

$$60 \exp\left(\frac{-20\,000}{RT}\right) = 1.3 \times 10^3 \exp\left(\frac{-40\,000}{RT}\right)$$

$$\exp\left(\frac{20\,000}{RT}\right) = 21.67$$

$$T = 782.1\text{K}$$

$$k_1 = 24.92 \text{ mol s}^{-1} \text{ m}^{-3} \text{ bar}^{-1}$$

$$k_2 = 13.85 \text{ mol s}^{-1}\text{m}^{-3} \text{ bar}^{-1}$$

$$k_3 = 9.69 \text{ mol s}^{-1} \text{ m}^{-3} \text{ bar}^{-1}$$

(b) $$n_A = n_{A0} - n_{A1} - n_{A2} - n_{A3}$$

$$n_B = n_{A1}, \qquad n_C = 2n_{A2}, \qquad n_D = 3n_{A3}$$

$$n_T = n_{A0} + n_{A2} + 2n_{A3}$$

$$Y_{B/A} = \frac{n_{A1}}{n_{A0}}, \ Y_{C/A} = \frac{n_{A2}}{n_{A0}}, \ Y_{D/A} = \frac{n_{A3}}{n_{A0}}, \ x_A = Y_{B/A} + Y_{C/A} + Y_{D/A}$$

$$\frac{n_{A2}}{n_{A1}} = \frac{k_2}{k_1}, \frac{n_{A3}}{n_{A1}} = \frac{k_3}{k_1}$$

$$n_T = n_{A0} + \frac{k_2}{k_1} n_{A1} + 2\frac{k_3}{k_1} n_{A1} \quad \text{and}$$

$$n_A = n_{A0} - n_{A1} - \frac{k_2}{k_1} n_{A1} - \frac{k_3}{k_1} n_{A1}$$

$$n_{A1} = \frac{k_1}{k_1 + k_2 + k_3}(n_{A0} - n_A)$$

$$n_T = n_{A0} + \frac{k_2 + 2k_3}{k_1 + k_2 + k_3}(n_{A0} - n_A)$$

$$\text{Let } a = \frac{k_2 + 2k_3}{k_1 + k_2 + k_3} = 0.686$$

$$V = -\int_0^{n_A} \frac{dn_A}{(k_1 + k_2 + k_3)\frac{n_A}{n_T}P}$$

$$= -\frac{1}{P(k_1 + k_2 + k_3)} \int_0^{n_A} \frac{n_{A0} + a(n_{A0} - n_A)}{n_A} dn_A$$

$$= -\frac{1}{P(k_1 + k_2 + k_3)} [(1 + a)n_{A0} \ln n_A - an_A]_{n_{A0}}^{n_A}$$

$$= -\frac{1}{P(k_1 + k_2 + k_3)} \left[(1 + a)n_{A0} \ln \frac{n_A}{n_{A0}} - an_A + an_{A0} \right]$$

$$= \frac{n_{A0}}{P(k_1 + k_2 + k_3)} \left[(1 + a) \ln \frac{1}{1 - x_A} - ax_A \right]$$

We require a 10% yield of D. As the selectivity for D is 20% we must react 50% of the A fed to the reactor,

$$Y_{D/A} = S_{D/A}x_A$$

$$x_A = 0.5$$

$$V = 0.034 \, \text{m}^3$$

4.1.2 CSTRs

Continuing with the same reaction scheme as previously (Reactions R4.4 and R4.5), the design equation for A will be,

$$r_A = \frac{n_{A0} - n_A}{V} = (k_1 + k_2)\frac{n_A}{v_T} \tag{4.25}$$

Rearranging, $\quad n_A = \dfrac{n_{A0}}{1 + (k_1 + k_2)\tau} \tag{4.26}$

To calculate n_R and n_S again we use the individual design equations for the two reactions,

$$r_{A1} = \frac{n_{A1}}{V} = k_1 \frac{n_A}{v_T} \tag{4.27}$$

$$r_{A2} = \frac{n_{A2}}{V} = k_2 \frac{n_A}{v_T} \tag{4.28}$$

where n_{A1} and n_{A2} are the molar flows of A that disappear through reactions 1 and 2 respectively. So,

$$\frac{n_{A1}}{n_{A2}} = \frac{k_1}{k_2} \tag{4.29}$$

Rearranging eqns 4.27 and 4.28,

$$n_{A1} = k_1 \tau n_A$$

$$n_{A2} = k_2 \tau n_A$$

$$n_R = n_{R0} + n_{A1} = n_{R0} + k_1 \tau n_A$$

$$n_S = n_{S0} + 2n_{A2} = n_{S0} + 2k_2 \tau n_A$$

Substituting for n_A from eqn 4.26,

$$n_R = n_{R0} + \frac{k_1 \tau n_{A0}}{1 + (k_1 + k_2)\tau}$$

$$n_S = n_{S0} + \frac{2k_2 \tau n_{A0}}{1 + (k_1 + k_2)\tau}$$

4.2   Parallel reactions of different order

4.2.1   Two reactions

Our treatment must be different when there are reactions of mixed reaction rate orders proceeding at the same time, e.g.

$$A \rightarrow R \quad r_{A1} = k_1 C_A \tag{R4.6}$$

$$2A \rightarrow S \quad r_{A2} = k_2 C_A^2 \tag{R4.7}$$

and the local or instantaneous selectivity depends upon the concentration of reactant,

$$\phi_{R/A} = \frac{r_{A1}}{r_{A1} + r_{A2}} = \frac{k_1 C_A}{k_1 C_A + k_2 C_A^2} \tag{4.30}$$

$$\phi_{S/A} = \frac{r_{A2}}{r_{A1} + r_{A2}} = \frac{k_2 C_A^2}{k_1 C_A + k_2 C_A^2} \tag{4.31}$$

In such a case, the choice of reactor will influence the overall selectivity. For a CSTR the overall selectivity is of course equal to the local selectivity as the local selectivity is a constant because of the lack of concentration gradients. Hence, calculation of the product distribution is straightforward. For a PFR or batch reactor it is more complex to evaluate the overall product distribution as the local selectivity or instantaneous selectivity will vary with position or time.

As we have two reactions we can write down two independent design equations. For a PFR, let us look at one design equation for both reactions combined and the other for Reaction R4.6 or reaction 1 alone,

$$r_A = -\frac{dn_A}{dV} \tag{4.32}$$

$$r_{A1} = \frac{dn_{A1}}{dV} \tag{4.33}$$

Integration of the first design equation is how we get a relationship between the outlet concentration of reactant and the volume of the reactor[7]. The second design equation cannot be integrated. There are three variables; the rate depends upon reactant concentration and we also have the n_{A1} and the reactor volume. We cannot relate n_{A1} to n_A because of the variable selectivity. To proceed to the product distribution we divide one design equation by the other (remember n_{A1} increases as n_A decreases), dividing eqn 4.33 by eqn 4.32,

$$\phi_{R/A} = -\frac{dn_{A1}}{dn_A} \tag{4.34}$$

This eliminates the volume variable and we recover the definition of local selectivity (compare to eqn 4.4).

The amount of A that disappears through reaction 1 is given by,

$$n_{A1} = \int_{inlet}^{exit} dn_{A1} = -\int_{n_{A0}}^{n_{Ae}} \phi_{R/A} dn_A \tag{4.35}$$

Therefore, the overall selectivity is given by,

$$S_{R/A} = \frac{n_{A1}}{n_{A0} - n_{Ae}} = \frac{-1}{(n_{A0} - n_{Ae})} \int_{n_{A0}}^{n_{Ae}} \phi_{R/A} dn_A \tag{4.36}$$

If the volumetric flow rate is a constant, then eqn 4.36 can be rewritten in terms of concentration,

$$S_{R/A} = \frac{-1}{(C_{A0} - C_{Ae})} \int_{C_{A0}}^{C_{Ae}} \phi_{R/A} dC_A \tag{4.37}$$

and the overall selectivity can be seen to be effectively an average of local selectivities over the whole reactor[8].

We can now compare the performance of CSTRs and PFRs (or batch reactors). In general, if the 'useful' reaction is of higher order than the 'wasteful' reaction then reactant concentrations should be kept as high as possible. This means that a PFR would perform better than a CSTR. If CSTRs must be used, then selectivity can be increased by using a number of CSTRs in series. If the volumes of the CSTRs are different then the smaller CSTRs should be placed first (large CSTRs early in the cascade would have the effect of diluting the reactant concentration early).

Consider what happens when the second-order reaction (Reaction R4.7 or reaction 2) is the desired reaction. The local selectivity for S is given by,

$$\phi_{S/A} = \frac{k_2 C_A^2}{k_1 C_A + k_2 C_A^2} = \frac{k_2 C_A}{k_1 + k_2 C_A}$$

[7] It is assumed that the reaction rates are only dependent on reactant concentrations and not product concentrations; if not, the differential equations will be coupled and they must be solved simultaneously.

[8] Remember that the minus sign is before the integral because we are integrating from the inlet to the outlet which corresponds to decreasing concentration of A and therefore the integral will have a negative value.

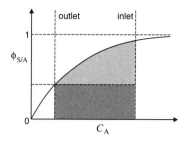

Fig. 4.2 Local selectivity for S versus concentration. The yield of a CSTR is proportional to the heavily shaded area; the yield of a PFR is proportional to the total shaded area.

[9] Unlike the previous case, this means that all reaction rates will be decreased. Process economics will dictate how far reaction rates can be reduced, and hence reactor volumes increased, in trying to improve selectivity.

Figure 4.2 shows a plot of local selectivity versus concentration. At zero concentration of A the selectivity for S is equal to zero. As concentration of reactant is increased the selectivity increases approaching unity for high concentrations. Because the CSTR operates at the outlet concentration of A, the yield of S depends only on the selectivity at that concentration and is therefore proportional to the heavily shaded rectangular area. The yield from the PFR involves integration of the local selectivities and from eqn 4.37 is proportional to the total shaded area under the curve. It can be seen that, for the same conversion, the yield from the PFR will be much better than for the CSTR.

Similarly, if the 'useful' reaction is of lower order than the 'wasteful' reaction then reactant concentrations should be kept as low as possible[9]. This means that a CSTR would perform better than a PFR. Consider the case where the first-order reaction (Reaction R4.6 or reaction 1) is desired. The local selectivity for R is given by,

$$\phi_{R/A} = \frac{k_1 C_A}{k_1 C_A + k_2 C_A^2} = \frac{k_1}{k_1 + k_2 C_A}$$

Figure 4.3 shows a plot of local selectivity versus concentration. At zero concentration of A the selectivity for R is equal to unity. As concentration of reactant is increased the selectivity decreases, approaching zero for high concentrations. The yield of R from the CSTR is proportional to the total shaded rectangular area. The yield from the PFR is proportional to the heavily shaded area under the curve. It can be seen that, for the same conversion, the yield from the CSTR will be much better than for the PFR.

Example 4.3 Consider two competing liquid-phase reactions:

$$A + B \rightarrow R \qquad r_R = k_1 \left(\frac{C_A}{C}\right) \left(\frac{C_B}{C}\right)^{-0.4}$$

$$A + B \rightarrow S \qquad r_S = k_2 \left(\frac{C_A}{C}\right)^{0.4} \left(\frac{C_B}{C}\right)^{1.2}$$

$$k_1 = 1.0 \, \text{mol litre}^{-1}\text{s}^{-1}, \quad k_2 = 2.0 \, \text{mol litre}^{-1}\text{s}^{-1};$$
$C = \text{reference concentration} = 1 \, \text{mol litre}^{-1}$;
$C_{A0} = C_{B0} = 1 \, \text{mol litre}^{-1}$

(The above rate data are expressed in terms of dimensionless concentrations. This is a common practice for empirical data as, otherwise, the units of the rate constants would become very inconvenient.)

What is the yield of R for: (a) a CSTR; (b) a PFR?

Both operate at 80% conversion. (If the initial concentrations of A and B were not in the stoichiometric ratio we would need to say whether the desired yield was defined in terms of consumption of A or B.)

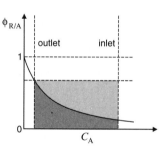

Fig. 4.3 Local selectivity for R versus concentration. The yield of a PFR is proportional to the heavily shaded area; the yield of a CSTR is proportional to the total shaded area.

Solution

$$\phi_{R/A} = \frac{r_{A1}}{r_{A1} + r_{A2}} = \frac{k_1 \left(\frac{C_A}{C}\right) \left(\frac{C_B}{C}\right)^{-0.4}}{k_1 \left(\frac{C_A}{C}\right) \left(\frac{C_B}{C}\right)^{-0.4} + k_2 \left(\frac{C_A}{C}\right)^{0.4} \left(\frac{C_B}{C}\right)^{1.2}}$$

$$= \frac{1}{1 + 2 \left(\frac{C_A}{C}\right)^{-0.6} \left(\frac{C_B}{C}\right)^{1.6}}$$

But, $C_A = C_B$; $\phi_{R/A} = \dfrac{1}{1 + 2\left(\frac{C_A}{C}\right)}$

(a) PFR

From eqn 4.37,

$$S_{R/A} = \frac{1}{\Delta C_A} \int_{C_{A0}}^{C_{Ae}} \phi_{R/A} \, dC_A = \frac{1}{\Delta C_A} \int_{C_{A0}}^{C_{Ae}} \frac{1}{1 + 2\left(\frac{C_A}{C}\right)} \, dC_A$$

where, $\Delta C_A = C_{Ae} - C_{A0}$

$$S_{R/A} = \frac{C}{2\Delta C_A} \left[\ln\left(1 + 2\frac{C_A}{C}\right) \right]_{C_{A0}}^{C_{Ae}} = \frac{1}{2(-0.8)} \ln \frac{1.4}{3} = 0.476$$

$$Y_{R/A} = 0.476 x_{Ae} = 38\%$$

(b) CSTR

$$S_{R/A} = \phi_{R/A} = \frac{1}{1 + 2\frac{C_{Ae}}{C}} = 0.714$$

$$Y_{R/A} = 0.714 x_{Ae} = 57\%$$

The CSTR is much more selective for reaction 1 because it works at the outlet concentration. Low concentrations relatively favour reaction 1 over reaction 2, because the overall reaction order is lower. Note that while a CSTR will give a better yield of reaction 1, it would need to be larger in volume than the PFR for a given conversion.

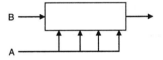

For bimolecular reactions, e.g.

$$\begin{aligned} A + B &\rightarrow C & r_{A1} &= k_1 C_A C_B \\ 2A &\rightarrow D & r_{A2} &= k_2 C_A^2 \end{aligned}$$

where the desired product is C, it is advantageous to keep the concentration of A low. This can be achieved by having a primary process stream containing B which is fed to the reactor inlet and then A is injected into the reactor or reactor chain as required (see Fig. 4.4). In this way the concentration of A can be kept low and the selectivity improved.

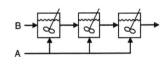

Fig. 4.4 Schematic representation of advantageous reactor systems for increasing selectivity when the desired reaction is bimolecular.

4.2.2 Three or more reactions

When three or more reactions are taking place in parallel and the desired reaction is of higher order than some of the competing reactions but of lower order than others, then an interesting problem is encountered. At low conversions the reactant concentrations are high and therefore the important competition may be between the desired reaction and reactions of higher orders. Therefore, concentrations should be reduced by using a CSTR first. At higher conversions, where reactant concentrations are lower, the important competition may be between the desired reaction and reactions of lower orders and now a PFR should be used to keep concentrations high. Therefore, it is possible that the best approach may involve a combination of CSTRs and PFRs.

Example 4.4   Consider three competing gas-phase reactions of different orders,

$$A \to B \qquad r_{A1} = k_1$$
$$A \to C \qquad r_{A2} = k_2 P_A$$
$$A \to D \qquad r_{A3} = k_3 P_A^2$$

with $k_1 = 1$ mol s^{-1} m^{-3}, $k_2 = 2$ mol s^{-1} m^{-3} bar^{-1}, $k_3 = 1$ mol s^{-1} m^{-3} bar^{-2}

The feed is pure A at a pressure of 5 bar. 96% conversion is required. What would the overall selectivity for C be if: (a) a PFR is used; (b) a CSTR is used; (c) the optimum reactor configuration to maximize the overall selectivity for C is used?

Solution   The local selectivity for C is given by,

$$\phi_{C/A} = \frac{k_2 P_A}{k_1 + k_2 P_A + k_3 P_A^2}$$

This function is plotted in Fig. 4.5(a). At a reactant partial pressure of zero the selectivity is zero, as the reactant partial pressure tends to infinity the selectivity tends to zero. For intermediate values of reactant partial pressure, a maximum in selectivity is obtained.

(a)   From eqn 4.36, for a PFR,

$$S_{C/A} = \frac{-1}{(P_{A0} - P_{Ae})} \int_{P_{A0}}^{P_{Ae}} \phi_{C/A} dP_A$$

$$= \frac{-1}{(P_{A0} - P_{Ae})} \int_{P_{A0}}^{P_{Ae}} \frac{k_2 P_A}{k_1 + k_2 P_A + k_3 P_A^2} dP_A$$

The overall selectivity for C is proportional to the area under the curve in Fig. 4.5(b).

Leaving P_A in bar and substituting for the rate constants,

$$\frac{k_2 P_A}{k_1 + k_2 P_A + k_3 P_A^2} = \frac{2 P_A}{1 + 2 P_A + P_A^2} = \frac{2 P_A}{(1 + P_A)^2} = \frac{2}{1 + P_A} - \frac{2}{(1 + P_A)^2}$$

$$S_{C/A} = \frac{-1}{(P_{A0} - P_{Ae})} \int_{P_{A0}}^{P_{Ae}} \left(\frac{2}{1 + P_A} - \frac{2}{(1 + P_A)^2} \right) dP_A$$

$$= \frac{-1}{(P_{A0} - P_{Ae})} \left[2 \ln(1 + P_A) + \frac{2}{(1 + P_A)} \right]_{P_{A0}}^{P_{Ae}}$$

$$= \frac{-1}{(4.8)} \left[2 \ln \frac{1.2}{6} + \frac{2}{1.2} - \frac{2}{6} \right] = 39.3\%$$

(b)   For a CSTR the overall selectivity is equal to the selectivity evaluated at the outlet partial pressure,

$$S_{C/A} = \phi_{C/A} = \frac{k_2 P_{Ae}}{k_1 + k_2 P_{Ae} + k_3 P_{Ae}^2} = \frac{2 P_{Ae}}{(1 + P_{Ae})^2} = \frac{0.4}{1.2^2} = 27.8\%$$

The overall selectivity for C is proportional to the area of the rectangle in Fig. 4.5(c).

(c)   At the inlet conditions, the selectivity is low compared to the maximum possible selectivity. Therefore, to optimize the overall selectivity, the partial pressure should be dropped to that corresponding to the occurrence in maximum local selectivity by using a CSTR.

The maximum in local selectivity occurs when, $\frac{d\phi_{C|A}}{dP_A} = 0$

Differentiating the expression for $\phi_{C/A}$,

$$\frac{d\phi_{C/A}}{dP_A} = \frac{k_2(k_1 + k_2 P_{A,max} + k_3 P_{A,max}^2) - k_2 P_{A,max}(k_2 + 2k_3 P_{A,max})}{(k_1 + k_2 P_{A,max} + k_3 P_{A,max}^2)^2} = 0$$

$$k_2(k_1 + k_2 P_{A,max} + k_3 P_{A,max}^2) = k_2 P_{A,max}(k_2 + 2k_3 P_{A,max})$$

$$P_{A,max} = \sqrt{\frac{k_1}{k_3}} = 1 \text{ bar}$$

The overall selectivity for the CSTR can then be evaluated,

$$S_{C/A,CSTR} = \frac{k_2 P_{A,max}}{k_1 + k_2 P_{A,max} + k_3 P_{A,max}^2} = 50\%$$

A PFR is now used to reduce the reactant partial pressure to its final value. After eqn 4.36,

$$S_{C/A,PFR} = \frac{-1}{(P_{A,max} - P_{Ae})}\left[2\ln(1 + P_A) + \frac{2}{(1 + P_A)}\right]_{P_{A,max}}^{P_{Ae}}$$

$$= \frac{-1}{(0.8)}\left[2\ln\frac{1.2}{2} + \frac{2}{1.2} - \frac{2}{2}\right] = 44.4\%$$

The overall selectivity for the combination of reactors is then an average based upon the amount of reactant converted in each reactor,

$$S_{C/A} = \frac{S_{C/A,CSTR}(P_{A0} - P_{A,max}) + S_{C/A,PFR}(P_{A,max} - P_{Ae})}{(P_{A0} - P_{Ae})}$$

$$= \frac{0.5(4) + 0.444(0.8)}{(4.8)} = 49.1\%$$

The shaded area in Fig. 4.5(d) is proportional to the optimum overall selectivity for the combination of a CSTR followed by a PFR.

4.3   Series reactions

Consider the two series reactions,

$$A \xrightarrow{k_1} R \xrightarrow{k_2} S$$

with both reactions first order.

$$r_A = k_1 C_A \tag{4.38}$$
$$r_R = k_1 C_A - k_2 C_R \tag{4.39}$$
$$r_S = k_2 C_R \tag{4.40}$$

where A is defined as a reactant (positive rates mean rates of disappearance) and both R and S are defined as products (positive rates mean rates of appearance).

4.3.1   Batch reactors and PFRs

Again, batch reactors and PFRs have similar behaviour and therefore we will only look at the behaviour of a batch reactor.

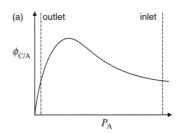

(a)

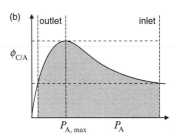

(b)

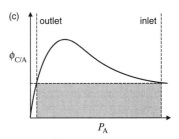

(c)

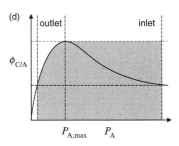

(d)

Fig. 4.5   (a) Local selectivity for C versus partial pressure (not to scale). (b) Overall selectivity of a PFR is proportional to shaded area. (c) Overall selectivity of a CSTR is proportional to shaded area. (d) Overall selectivity of optimum reactor configuration (CSTR, with outlet at the composition corresponding to maximum selectivity, followed by PFR) is proportional to shaded area.

Solving the design equation for the amount of A present in a batch reactor (see eqn 3.5), we get, as before,

$$N_A = N_{A0}\, e^{-k_1 t} \tag{4.41}$$

We can write down the differential equation or design equation governing the number of moles of R present using eqn 4.39,

$$r_R = k_1 C_A - k_2 C_R = \frac{1}{V}\frac{dN_R}{dt} \tag{4.42}$$

$$\frac{1}{V}\frac{dN_R}{dt} = k_1 \frac{N_A}{V} - k_2 \frac{N_R}{V} \tag{4.43}$$

and substituting in for the time dependence of the number of moles of A (eqn 4.41),

$$\frac{dN_R}{dt} + k_2 N_R = k_1 N_{A0}\, e^{-k_1 t} \tag{4.44}$$

We now try a solution of the appropriate form and evaluate the coefficients,

$$N_R = A e^{-k_1 t} + B e^{-k_2 t}$$

and evaluate A and B by substitution into the original differential equation with the initial condition,

$$t = 0, N_R = 0$$

This gives[10],

$$N_R = \frac{k_1 N_{A0}}{k_2 - k_1}\left\{ e^{-k_1 t} - e^{-k_2 t} \right\} \tag{4.45}$$

[10] Try substituting this back in the original differential equation to show that it is a solution.

We can find the number of moles of S present because the total number of moles is conserved,

$$N_S = N_{A0} - N_A - N_R$$

$$\frac{N_S}{N_{A0}} = 1 - e^{-k_1 t} - \frac{k_1}{k_2 - k_1}\left\{ e^{-k_1 t} - e^{-k_2 t} \right\}$$

In many cases R may be a valuable product and hence we may want to maximize its production. The optimum residence time to maximize the yield of R can be easily found using eqn 4.45,

$$\frac{N_R}{N_{A0}} = \frac{k_1}{k_2 - k_1}\left\{ e^{-k_1 t} - e^{-k_2 t} \right\}$$

and to maximize N_R, we need,

$$\frac{dN_R}{dt} = 0 = \frac{k_1 N_{A0}}{k_2 - k_1}\left\{ -k_1\, e^{-k_1 t_{opt}} + k_2\, e^{-k_2 t_{opt}} \right\}$$

$$k_1\, e^{-k_1 t_{opt}} = k_2\, e^{-k_2 t_{opt}}$$

$$t_{opt} = \frac{\ln\frac{k_2}{k_1}}{k_2 - k_1} = \frac{1}{k_{lm}} \tag{4.46}$$

where k_{lm} is the log-mean rate constant. Substitution of the expression for optimum residence time (eqn 4.46) into eqn 4.41 will give us an expression for the maximum yield of R which can be obtained in a batch reactor.

4.3.2 CSTRs

Solving the design equation for A (see eqn 3.16),

$$n_A = \frac{n_{A0}}{1 + k_1 \tau} \tag{4.47}$$

The design equation for R,

$$\frac{n_R - n_{R0}}{V} = k_1 \frac{n_A}{v_T} - k_2 \frac{n_R}{v_T} \tag{4.48}$$

For no R initially present,

$$n_R = k_1 \tau n_A - k_2 \tau n_R \tag{4.49}$$

and rearranging and substituting eqn 4.47 into eqn 4.49,

$$n_R = \frac{k_1 \tau n_A}{1 + k_2 \tau} = \frac{k_1 \tau n_{A0}}{(1 + k_1 \tau)(1 + k_2 \tau)} \tag{4.50}$$

In addition $\quad n_S = n_{A0} - n_A - n_R$

$$\frac{n_S}{n_{A0}} = 1 - \frac{1}{1 + k_1 \tau} - \frac{k_1 \tau}{(1 + k_1 \tau)(1 + k_2 \tau)}$$

$$= \frac{k_1 k_2 \tau^2}{(1 + k_1 \tau)(1 + k_2 \tau)}$$

For the optimum residence time to maximize the yield of R we differentiate eqn 4.50 with respect to residence time,

$$\frac{dn_R}{d\tau} = 0$$

$$\frac{d}{d\tau} \frac{k_1 k_2 \tau^2}{(1 + k_1 \tau)(1 + k_2 \tau)} = 0$$

Differentiating and manipulating (see Problem 4.5 at the end of the chapter) gives,

$$\tau_{opt} = \frac{1}{\sqrt{k_1 k_2}} = \frac{1}{k_{gm}} \tag{4.51}$$

where k_{gm} is the geometric-mean rate constant.

4.3.3 Comparison of PFRs and CSTRs

Let us consider what will happen if we have a PFR and a CSTR both operating at the same residence time (see Fig. 4.6). If we have a positive-order reaction, the outlet concentration of reactant from the PFR will be lower than that of the CSTR because of the lower rates exhibited by the CSTR as the reactant is immediately diluted. This means that, at short residence times, the outlet concentration of the intermediate R will be higher from the PFR because of its greater rate of production. This gives the appearance that things happen quicker in the PFR than in the CSTR. A maximum in outlet concentration of R is reached in the PFR at a shorter residence time than that required to reach a maximum in the CSTR. Most importantly, the maximum outlet concentration of R from the PFR is greater than the maximum outlet concentration of R from the CSTR. This is because when we have a series reaction, if we want to maximize the outlet concentration of an intermediate, we want to make sure

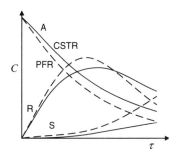

Fig. 4.6 Concentrations (not to scale) of species involved in a series reaction, $A \rightarrow R \rightarrow S$, in a PFR and a CSTR.

that all molecules react for the same length of time (this time being the optimum residence time). This is what happens in a PFR—all molecules have exactly the same residence time. However, in a CSTR not all molecules have the same residence time. The residence time of a CSTR is a mean residence time, some molecules pass through quickly and some pass through the reactor more slowly. As a result the maximum yield of an intermediate can never be as good from a CSTR as it would be from a PFR (or batch reactor).

4.4   Problems

4.1 The following liquid-phase reactions were performed in a PFR,

$$\begin{array}{ll} A \rightarrow 2R & r_{A1} = k_1 C_A \\ A \rightarrow 3S & r_{A2} = k_2 C_A \end{array}$$

For a residence time of 40 s the conversion of A was found to be 60% and the ratio of moles of R produced to moles of S produced was found to be 4 to 1. The reactor feed was pure A. Calculate the values of k_1 and k_2.

4.2 The following reactions are carried out isothermally in the gas phase in a PFR:

$$\begin{array}{ll} A \rightarrow 2R & r_{A1} = k_1 P_A \\ A \rightarrow 3S & r_{A2} = k_2 P_A \end{array}$$

$P = 1$ bar, $k_1 = 20 \, \text{mol s}^{-1} \text{m}^{-3} \text{bar}^{-1}$; $k_2 = 40 \, \text{mol s}^{-1} \text{m}^{-3} \text{bar}^{-1}$
Pure A is fed to the reactor at 1 mol s^{-1}. What reactor volume is required for a 30% yield of R?

4.3 Consider the following reactions,

$$\begin{array}{ll} A \rightarrow B, & r_{A1} = k_1 C_A \\ 2A \rightarrow C, & r_{A2} = k_2 C_A^2 \end{array}$$

B is the desired product and C is a waste product.
$k_1 = 1 \, \text{s}^{-1}$, $k_2 = 10 \, \text{litre mol}^{-1} \text{s}^{-1}$,
$v_T = \text{const} = 1 \, \text{litre s}^{-1}$,
$C_{A0} = 1 \, \text{mol litre}^{-1}$
(a) What is the volume of a PFR required for 95% conversion of A?
(b) What is the yield of B?
(c) What is the overall selectivity for B?
(d) What is the conversion of A in a CSTR of the same volume?
(e) What is the yield of B in this case?
(f) What is the overall selectivity for B?
(g) What would be the yield of B if the volume of the CSTR were increased to give 95% conversion?
(h) What is the overall selectivity for B?

4.4 Two competing reactions, one of which is autocatalytic, are performed in a CSTR,

$$\begin{array}{ll} A \rightarrow B & r_{A1} = k_1 C_A C_B \\ A \rightarrow C & r_{A2} = k_2 C_A \end{array}$$

where r_{A1} is the rate of disappearance of A through reaction 1, etc. There is no B present in the feed to the reactor. (a) Given that reaction 1 is the desired reaction, indicate how you would start up the reactor.
(b) Show that, at steady state,

$$C_B = C_{A0} - \frac{1}{\tau k_1} - \frac{k_2}{k_1}$$
$$\text{for} \quad C_{A0} > \frac{1}{\tau k_1} + \frac{k_2}{k_1}$$

(c) If the reactor is operated at a conversion such as to maximize the rate of reaction 1, show that at steady state the selectivity for B will be,

$$S_{B/A} = \frac{1 - \frac{k_2}{k_1 C_{A0}}}{1 + \frac{k_2}{k_1 C_{A0}}}$$

4.5 The intermediate R is to be produced in a continuous reactor:

$$A \xrightarrow{k_1} R \xrightarrow{k_2} S \quad k_1 = 2 \, \text{s}^{-1} \text{ and } k_2 = 0.5 \, \text{s}^{-1}$$

Calculate the value of τ for a CSTR and a PFR for maximum production of R. For this optimum value of τ, calculate the conversion of A and the yields of R and S.

4.6 Consider the reaction network shown in Fig. 4.7. All reactions are first order with respect to the reactant indicated.

$$k_3/k_1 = 0.4; \quad (k_2 + k_4)/k_1 = 0.2$$

Determine the yield of B when the conversion of A is 70% in: (a) a PFR; (b) a CSTR.

Fig. 4.7 Reaction network for Problem 4.6.

5 The energy balance and temperature effects

We will now no longer confine ourselves to considering isothermal reactors. Let us start by looking at the temperature dependence of reaction kinetics.

5.1 Temperature dependence of reaction rate

5.1.1 Irreversible reaction

For a first-order irreversible reaction,

$$r_A = kC_A$$

Substituting in the temperature dependence for the rate constant and using conversion,

$$r_A = k' \exp\left(-\frac{\Delta E}{RT}\right) C_{A0}(1 - x_A)$$

where k' is the pre-exponential constant and ΔE is the activation energy of the reaction. Figure 5.1 shows how the rate of reaction varies as a function of temperature and conversion (lines of constant rate are shown). As temperature is increased reaction rate increases; as conversion is decreased reaction rate increases as the concentration of reactant is higher. A very low rate will be apparent either at a very low temperature or a very high conversion (approaching unity). In fact, a line of zero rate would correspond to the y-axis (zero temperature) and the line at a conversion of unity. As we move away from this envelope, reaction rate increases. Clearly, for any given conversion, we would wish to operate at the highest possible temperature (this will be limited by factors such as materials of construction) to maximize the reaction rate and therefore minimize the reactor size.

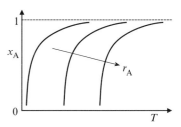

Fig. 5.1 Lines of constant rate of reaction shown in conversion–temperature space for a first-order irreversible reaction.

5.1.2 Reversible endothermic reaction

Figure 5.2 shows an energy profile for an endothermic reaction, A ⇔ B. The reactants are on the left-hand side and the products on the right-hand side. The energy associated with the products is greater than that associated with the reactants for an endothermic reaction. Therefore, the forward activation energy, ΔE_1, must be greater than the reverse activation energy, ΔE_{-1}[1]. The net reaction rate is given by the difference in the forward and reverse reaction rates,

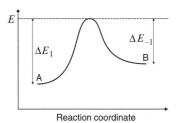

Fig. 5.2 Energy profile for the endothermic reaction, A ⇔ B.

[1] The difference between the activation energies is equal to the change in internal energy which, for a reaction with an equal number of moles on both sides, is equal to the heat of reaction,

$$\Delta E_1 - \Delta E_{-1} = \Delta U = \Delta H_R$$

$$r_A = k_1 C_A - k_{-1} C_B$$

For no B present initially,

$$r_A = k_1' \exp\left(-\frac{\Delta E_1}{RT}\right) C_{A0}(1 - x_A) - k_{-1}' \exp\left(-\frac{\Delta E_{-1}}{RT}\right) C_{A0} x_A$$

At equilibrium, $r_A = 0$; $x_A = x_A^*$

Substituting into the reaction rate expression,

$$\frac{x_A^*}{1 - x_A^*} = \frac{k_1}{k_{-1}} = K = \frac{k_1'}{k_{-1}'} \exp\left(\frac{\Delta E_{-1} - \Delta E_1}{RT}\right)$$

where K is the equilibrium constant.

Therefore, $x_A^* = \dfrac{K}{1 + K}$

As temperature increases, the value of K increases for an endothermic reaction and therefore the value of the equilibrium conversion, x_A^*, increases. Figure 5.3 shows the form of the equilibrium line for a reversible endothermic reaction. High equilibrium conversion are favoured at high temperatures. The equilibrium line is a line of constant reaction rate (this rate of course being equal to zero). Other lines of constant rate are shown in Fig. 5.3. Low rates are obtained close to the equilibrium line. Moving away from the equilibrium line serves to increase reaction rate. As in the case of the irreversible reaction, we would wish to operate at the highest possible temperature to maximize reaction rate.

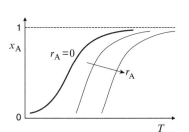

Fig. 5.3 Lines of constant rate of reaction shown in conversion–temperature space for a reversible endothermic reaction.

5.1.3   Reversible exothermic reaction

Again, $x_A^* = \dfrac{K}{1 + K}$

The forward activation energy, ΔE_1, is now less than the reverse activation energy, ΔE_{-1}. As temperature increases, the equilibrium constant will decrease and therefore the equilibrium conversion will decrease. Figure 5.4(a) shows the equilibrium line in conversion–temperature space. Once more this line corresponds to the line of zero reaction rate. However, there is also zero reaction rate at zero temperature (the y-axis). Again, rate is increased if we move away from the zero rate envelope.

Let us now consider a line of constant conversion. If we proceed along this line from low temperature (at a temperature of absolute zero the rate will be zero—point A) the rate will increase. However, at some point it will just touch one of the lines of constant reaction rate (point B—and this will correspond to the highest rate achievable at this conversion) and then as temperature increases further the rate will drop reaching zero at the equilibrium line (point C). The rate at constant conversion can be plotted against temperature and this is illustrated in Fig. 5.4(b).

This means that if we have a reactor operating at a particular conversion there will be an optimum temperature that will give us a maximum in reaction rate. It is important to be able to calculate this optimum temperature.

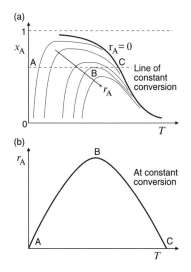

Fig. 5.4   (a) Lines of constant rate of reaction shown in conversion–temperature space for a reversible exothermic reaction. Also shown is a line at constant conversion. On this line the rate will be zero at point A, will reach a maximum at point B, and will be zero at point C. (b) Plot of rate versus temperature on line of constant conversion.

For the reaction A $\Leftrightarrow$ B with first-order kinetics,

$$r_A = k_1 C_A - k_{-1} C_B$$

For no B present initially,

$$r_A = k_1 C_{A0}(1 - x_A) - k_{-1} C_{A0} x_A$$
$$r_A = C_{A0}\{k_1 - (k_1 + k_{-1})x_A\}$$

The maximum reaction rate for a given conversion will occur when,

$$\frac{dr_A}{dT} = 0 \tag{5.1}$$

Differentiating the kinetic expression, with C_{A0} and x_A as constants,

$$\frac{dr_A}{dT} = C_{A0}\left\{\frac{dk_1}{dT} - \left(\frac{dk_1}{dT} + \frac{dk_{-1}}{dT}\right)x_A\right\} = 0$$

$$x_A = \frac{\dfrac{dk_1}{dT}}{\dfrac{dk_1}{dT} + \dfrac{dk_{-1}}{dT}} = \frac{1}{1 + \dfrac{dk_{-1}}{dT}\bigg/\dfrac{dk_1}{dT}}$$

But, in general, $k = k' \exp\left(-\dfrac{\Delta E}{RT}\right)$

Therefore,

$$\frac{dk_1}{dT} = k_1' \frac{\Delta E_1}{RT^2} \exp\left(-\frac{\Delta E_1}{RT}\right) = k_1 \frac{\Delta E_1}{RT^2}$$
$$\frac{dk_{-1}}{dT} = k_{-1}' \frac{\Delta E_{-1}}{RT^2} \exp\left(-\frac{\Delta E_{-1}}{RT}\right) = k_{-1} \frac{\Delta E_{-1}}{RT^2}$$

Therefore,

$$x_A = \frac{1}{1 + \dfrac{\Delta E_{-1} k_{-1}}{\Delta E_1 k_1}} = \frac{1}{1 + \dfrac{\Delta E_{-1}}{\Delta E_1}\dfrac{1}{K}} \tag{5.2}$$

This means that for any conversion the temperature that results in the maximum reaction rate can be easily found from eqn 5.2 (or its equivalent if a different kinetic relationship is obeyed). Therefore, as a CSTR operates at only one conversion, it should be operated isothermally at the appropriate temperature to maximize rate or minimize its volume. However, conversion will vary along the length of a PFR and therefore the optimum temperature will also vary. This results in an optimum operating line (in conversion–temperature space) for a PFR[2].

[2] Conversion will vary with time in a batch reactor and, therefore, to operate at the minimum necessary residence time to reach a given final conversion, the temperature of the reactor will need to change with time in the appropriate manner.

Example 5.1

The exothermic gas phase reaction,

$$A \Leftrightarrow B + C; \quad r_A = k_1 P_A - k_{-1} P_B P_C$$

is to be carried out in a CSTR operating at 50 bar.
Calculate the maximum rate if the reactor operates at a 22% conversion of A (feed is pure A).

$$k_1 = 0.435 \exp\left(\frac{-20\,000}{RT}\right) \text{mol s}^{-1} \text{ m}^{-3} \text{ bar}^{-1}$$

$$k_{-1} = 147 \exp\left(\frac{-60\,000}{RT}\right) \text{mol s}^{-1} \text{ m}^{-3} \text{ bar}^{-2}$$

RT is in J mol^{-1}; $R = 8.314$ J mol^{-1} K^{-1}

Solution

$$A \Leftrightarrow B + C$$

$$n_A = n_{A0} - n_{A0}x_A, \quad n_B = n_{A0}x_A, \quad n_C = n_{A0}x_A, \quad n_T = n_{A0} + n_{A0} + n_{a0}x_A$$

$$P_A = \frac{n_A}{n_T} P = \frac{1 - x_A}{1 + x_A} P = 31.97 \text{ bar}$$

$$P_B = P_C = \frac{x_A}{1 + x_A} P = 9.02 \text{ bar}$$

$$r_A = k_1 P_A - k_{-1} P_B P_C$$

$$\frac{dr_A}{dT} = \frac{dk_1}{dT} P_A - \frac{dk_{-1}}{dT} P_B P_C = 0$$

$$\frac{dk_1}{dT} \bigg/ \frac{dk_{-1}}{dT} = \frac{P_B P_C}{P_A}$$

$$k_1 = k_1' \exp\left(\frac{-\Delta E_1}{RT}\right); \quad k_{-1} = k_{-1}' \exp\left(\frac{-\Delta E_{-1}}{RT}\right)$$

$$\frac{dk_1}{dT} \bigg/ \frac{dk_{-1}}{dT} = \frac{\Delta E_1 k_1}{\Delta E_{-1} k_{-1}} = \frac{P_B P_C}{P_A}$$

$$\frac{k_1}{k_{-1}} = 3.05 \times 10^{-3} \exp\left(\frac{40\,000}{RT}\right) = \frac{\Delta E_{-1} P_B P_C}{\Delta E_1 P_A} = \frac{60}{20} \frac{(9.02)^2}{31.97} \text{bar} = 7.63 \text{ bar}$$

$$T = 614.8 \text{ K}$$

$$k_1 = 8.69 \times 10^{-3} \text{mol s}^{-1} \text{ m}^{-3} \text{ bar}^{-1}$$

$$k_{-1} = 1.17 \times 10^{-3} \text{mol s}^{-1} \text{ m}^{-3} \text{ bar}^{-2}$$

$$r_A = 0.278 - 0.095 = 0.183 \text{ mol s}^{-1} \text{ m}^{-3}$$

5.2   The energy balance

Up until now we have only used the material balance in reactor design. Now, if we want to take account of temperature effects, we must also include an energy balance.

5.2.1   CSTRs

Consider a process stream at an initial temperature of T_0 with a conversion defined as zero. This process stream is now passed through a reactor and at some later time is at a temperature T and conversion x_A. Because energy is an intensive property it is not important what path the stream followed in conversion–temperature space from one point to the other so we will consider a path that makes the formulation of the energy balance straightforward (see Fig. 5.5). First, we will heat the mixture up while at constant composition, and then we will allow the mixture to react at constant temperature.

First, let us consider the energy involved in heating the process stream at constant initial composition,

$$Q_r = n_{A0}\overline{c_{PA}}(T - T_0) + n_{B0}\overline{c_{PB}}(T - T_0) + \cdots$$

where $\overline{c_{Pi}}$ is the mean (mean because it is an average over an appropriate temperature range) specific heat capacity of the ith component (with units, for example, J mol^{-1} K^{-1}). Q_r is the heat required to heat up the process stream or alternatively the heat removed by the process stream (in, for example, J s^{-1}),

$$Q_r = \sum_i n_{i0}\overline{c_{Pi}}(T - T_0) \tag{5.3}$$

Defining an overall mean specific heat capacity as,

$$\overline{c_{P0}} = \sum_i \frac{n_{i0}}{n_{T0}}\overline{c_{Pi}}$$

$$Q_r = n_{T0}\overline{c_{P0}}(T - T_0) \tag{5.4}$$

Now let us consider the energy involved with the change in composition due to reaction,

$$Q_g = n_{A0}x_A(-\Delta H_R(T)) \tag{5.5}$$

where Q_g is the heat generated by the reaction. It should be remembered that the heat of reaction will, in general, be a function of temperature.

The energy balance is of the form,

$$Q_g = Q_r + Q \tag{5.6}$$

i.e. the heat generated by the reaction goes in heating up the process stream and also supplies any heat, Q, that is removed from the system. Therefore, substituting eqns 5.4 and 5.5 into eqn 5.6,

$$n_{A0}x_A(-\Delta H_R(T)) = n_{T0}\overline{c_{P0}}(T - T_0) + Q \tag{5.7}$$

This is then a general form of the energy balance.

We will now make some simplifying assumptions that will make the energy balance easier to use.

1. We will assume that the heat of reaction is independent of temperature, i.e.

$$-\Delta H_R(T) = -\Delta H_R$$

2. We will assume that the heat capacity of the stream is a constant, i.e. it is not function of composition[3],

$$n_{T0}\overline{c_{P0}} = n_T\overline{c_P}$$

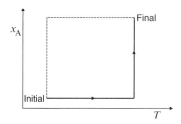

Fig. 5.5 Initial and final states of a process stream in conversion–temperature space.

[3] To illustrate this assumption let us consider the reaction of hydrogen with iodine at 300 K,

$$H_2 + I_2 \Leftrightarrow 2HI$$

The specific heat capacities of the individual components are as follows,

$c_{PH_2} = 28.7$ J mol^{-1} K^{-1}

$c_{PI_2} = 37.6$ J mol^{-1} K^{-1}

$c_{PHI} = 30.0$ J mol^{-1} K^{-1}

The heat capacity of one mole of hydrogen and one mole of iodine is then equal to 66.3 J K^{-1}. This compares to the heat capacity of two moles of hydrogen iodide of 60.0 J K^{-1}. It can be seen that there is only a difference of about 10% in the two values, hence the assumption that total heat capacity remains constant is reasonable.

However, it must be remembered that n_T and $\overline{c_P}$ must be evaluated at the same composition.

With these two assumptions we can simplify the energy balance expression,

$$n_{A0}x_A(-\Delta H_R) = n_T\overline{c_P}(T - T_0) + Q \tag{5.8}$$

For an adiabatic reactor where no heat is added to, or removed from, the process stream, i.e. $Q = 0$,

$$n_{A0}x_A(-\Delta H_R) = n_T\overline{c_P}(T - T_0) \tag{5.9}$$

Example 5.2   A liquid-phase reaction is performed in a CSTR,

A $\rightarrow$ Products

(a)   What is the operating temperature if the reactor is adiabatic?

(b)   What is the operating temperature if the reactor is surrounded by a cooling jacket with coolant at 300 K? The heat transfer area, A, is 0.2 m^2 and the overall heat transfer coefficient, U, is 500 W m^{-2} K^{-1}.

Inlet concentration, C_{A0}	10 mol litre^{-1}
Volumetric flow rate, v_T	0.1 litre s^{-1}
Conversion, x_A	80%
Feed temperature, T_O	300 K
Heat of reaction, ΔH_R	-100 kJ (mol of A)$^{-1}$
Overall mean heat capacity, $\overline{c_P}$	4.2 kJ litre^{-1} K^{-1}

Solution   (a) $Q_g = n_{A0}x_A(-\Delta H_R) = v_T C_{A0}x_A(-\Delta H_R)$

$$Q_r = v_T\overline{c_P}(T - T_0)$$

(Note: the units of the mean heat capacity imply that it is the volumetric flow rate of the stream that should be used and not the molar flow rate—remember, always check the units!)

$$Q_g = Q_r$$

$$\Delta T = \frac{C_{A0}x_A(-\Delta H_R)}{\overline{c_P}} = \frac{(10)(0.8)(10^5)}{4200} = 190 \text{ K}$$

$$T = 490 \text{ K}$$

(b) $Q_g = v_T C_{A0}x_A(-\Delta H_R)$

$$Q_r = v_T\overline{c_P}(T - T_0)$$

$$Q = UA(T - T_j)$$

$$Q_g = Q_r + Q$$

$$\Delta T = \frac{v_T C_{A0}x_A(-\Delta H_R)}{v_T\overline{c_P} + UA} = \frac{(0.1)(10)(0.8)(10^5)}{(0.1)(4200) + (500)(0.2)} = 154 \text{ K}$$

$$T = 454 \text{ K}$$

5.2.2   PFRs

For PFRs we need to use an energy balance that is in differential form. Considering a differential change in temperature caused by a differential change in composition with a differential amount of heat removed from the system, dQ, and using a similar approach as for the derivation of eqn 5.8,

$$n_{A0}dx_A(-\Delta H_R) = n_T\overline{c_P}dT + dQ \tag{5.10}$$

If q is the heat flux through the wall of a PFR, eqn 5.10 becomes,

$$n_{A0}dx_A(-\Delta H_R) = n_T\overline{c_P}dT + 2\pi r dlq \tag{5.11}$$

as $2\pi r dl$ is the differential area of a tubular reactor wall associated with an element of length dl.

Remembering that the material balance for a PFR (eqns 2.8 and 2.12) can be written as,

$$r_A dV = -dn_A = n_{A0}dx_A \tag{5.12}$$

Substituting eqn 5.12 into the energy balance (eqn 5.11) gives,

$$r_A dV(-\Delta H_R) = n_T\overline{c_P}dT + 2\pi r dlq$$

$$\text{But,} \quad dV = \pi r^2 dl$$

$$r_A(-\Delta H_R) = n_T\overline{c_P}\frac{dT}{dV} + \frac{2q}{r}$$

$$\frac{dT}{dV} = \frac{r_A(-\Delta H_R) - \frac{2q}{r}}{n_T\overline{c_p}} \tag{5.13}$$

which determines the temperature profile along the reactor. The material balance (eqn 5.12), of course, determines the composition profile along the reactor,

$$\frac{dx_A}{dV} = \frac{r_A}{n_{A0}} \tag{5.14}$$

If we divide eqn 5.14 by eqn 5.13,

$$\frac{dx_A}{dT} = \frac{n_T\overline{c_P}}{n_{A0}(-\Delta H_R) - \frac{n_{A0}}{r_A}\frac{2q}{r}} \tag{5.15}$$

Equation 5.15 relates the conversion to the temperature and can therefore be considered to describe an operating line for the PFR[4].

The heat flux can take different forms. In the adiabatic case, eqn 5.15 becomes,

$$\frac{dx_A}{dT} = \frac{n_T\overline{c_P}}{n_{A0}(-\Delta H_R)} \tag{5.16}$$

and the gradient of the operating line is constant. Figure 5.6 shows such straight adiabatic operating lines for different heats of reaction. The gradient of the operating line is small (and positive for an exothermic reaction or negative for an endothermic reaction) if the reactant is at high concentration or the heat of reaction is large. This small gradient means that small changes in composition will result in large changes in temperature.

For electrical heating, the heat flux is constant,

$$q = q_0 \tag{5.17}$$

whereas with a cooling or heating medium the heat flux depends upon the difference in temperature between the process stream and jacket,

$$q = U(T - T_j) \tag{5.18}$$

Figure 5.7 shows the operating lines for an exothermic reaction with heat input and heat removal (heating and cooling with a heating/cooling

[4] In the case of a CSTR the concept of an operating line is meaningless; instead, the operating point of the CSTR is dictated by the simultaneous solution of the material and energy balances, i.e. the intersection of the material balance line and the energy balance line in conversion–temperature space.

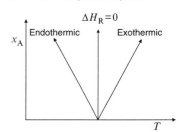

Fig. 5.6 Operating lines for an adiabatic reactor in conversion–temperature space for endothermic and exothermic reactions.

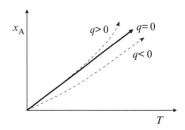

Fig. 5.7 Operating line (from the energy balance) for an exothermic reaction with adiabatic operation ($q = 0$) and operation with heat input ($q < 0$) and heat removal ($q > 0$).

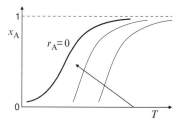

Fig. 5.8 Adiabatic operating line for a reversible endothermic reaction.

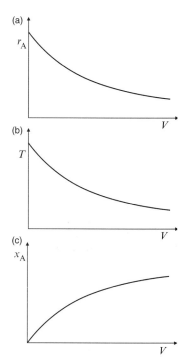

Fig. 5.9 (a) Rate of reaction, (b) temperature, and (c) conversion as a function of axial position (or reactor volume) for a reversible endothermic reaction.

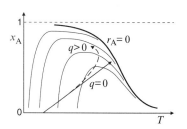

Fig. 5.10 Operating line for a reversible exothermic reaction for adiabatic operation ($q = 0$) and operation with heat removal ($q > 0$).

medium) and adiabatic operation. In the adiabatic case the operating line is a straight line. In the case of heat input, the temperature rises more quickly than the adiabatic case (this is particularly important at low process stream temperatures where heating is greatest). In the case of heat removal, the temperature rises more slowly than in the adiabatic case (this is particularly important at high process stream temperatures).

Note that the equations dictating the temperature profile and composition profile (eqns 5.13 and 5.14) are not independent but must be solved together as coupled differential equations (the rate of reaction is a function of both composition and temperature). The solution, in general, must be performed numerically. However, let us consider some general forms of the solution.

Reversible endothermic reaction. Figure 5.8 shows an adiabatic operating line in conversion–temperature space (lines of constant rate are also shown). The rate will monotonically decrease from the inlet (Fig. 5.9(a)) as will the temperature (Fig. 5.9(b)). The conversion profile is shown in Fig. 5.9(c). The gradients of both the temperature and conversion profiles are directly related to the reaction rate (through eqns 5.13 and 5.14) and at lower rates (i.e. further along the reactor) the gradients tend to be lower.

Reversible exothermic reaction. Let us consider the operating line for a reversible exothermic reaction. Figure 5.10 shows the operating line for the adiabatic case ($q = 0$) as well as for the case of heat removal ($q > 0$). Initially, the rate will increase along the reactor due to the increasing temperature of the process stream and then, as equilibrium is approached, the rate will begin to drop. Figure 5.11 shows the general forms of the reaction rate, temperature, and conversion profiles.

For the adiabatic case, the temperature and conversion profiles (Figs 5.11(b) and 5.11(c)) have a similar shape as the temperature must be linearly dependent upon the conversion (from integration of eqn 5.15). At the reactor inlet the rate of reaction is low and therefore the gradients of the temperature and conversion profiles are low. As the temperature rises the rate increases and finally begins to fall as equilibrium is approached. This results in the characteristic sigmoidal shape of the temperature and conversion profiles. For heat addition ($q < 0$) the solution is not changed significantly from the adiabatic case ($q = 0$); however, the case of heat removal ($q > 0$) is of interest.

In the case of heat removal, as the rate of reaction begins to fall, the rate of heat generation will decrease and will eventually become equal to the rate of heat removal. At this point the gradient of the operating line will be infinite (Fig. 5.10). Further along the reactor still, the rate of heat removal will be greater than the rate of heat generation and the temperature of the process stream will fall (Fig. 5.11(b)). This results in a 'hot-spot' in the reactor. This heat removal takes the process stream away from the equilibrium condition and therefore allows higher conversions to be achieved in the reactor. Heat removal will lower rates of reaction early in the reactor, but, at high conversions, heat removal will keep the stream further from equilibrium and therefore result in higher reaction rates (Fig. 5.11(a)).

Adiabatic reactor. As previously mentioned, in general, the reactor volume for a PFR cannot be calculated analytically because of the coupled material and energy balances. However, in the adiabatic case, the equation for the operating line can be determined analytically from integration of eqn 5.16 (this effectively uncouples eqns 5.13 and 5.14). As we know the relationship between temperature and conversion on the operating line, the reaction rate can be evaluated at any point and a numerical integration of eqn 5.14 performed. If n_{A0}/r_A can be plotted against conversion (see Fig. 5.12), the PFR volume is simply equal to the area under the curve. One simple method of numerical integration, or estimating the area under the curve, is the trapezoidal rule (see Fig. 5.13). The function to be integrated, n_{A0}/r_A, is evaluated at a number of values of conversion. The approximate area under the curve is then given by summing the areas of all of the individual elements. For instance, the area enclosed between two function evaluations is given approximately by,

$$\text{area} = \frac{(f_{i-1} + f_i)}{2} \Delta x_A$$

If all function evaluations are performed at uniform spacing, i.e. Δx_A is constant, then the approximate area under the curve will be given by,

$$\int_0^{x_{Ae}} f \, dx_A \approx \Delta x_A \left(\frac{f_0}{2} + f_1 + \cdots f_{n-1} + \frac{f_n}{2} \right)$$

where, $\Delta x_A = x_{Ae}/n$, as the range of integration is broken down into n equal subdivisions. As the value of n increases the accuracy of the estimate will increase.

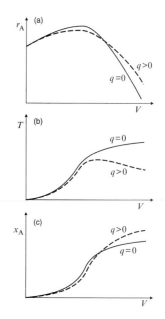

Fig. 5.11   (a) Rate of reaction, (b) temperature, for a reversible exothermic reaction and (c) conversion as a function of axial position (or reactor volume) for a reversible exothermic reaction for adiabatic operation ($q = 0$) and operation with heat removal ($q > 0$).

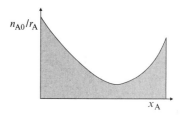

Fig. 5.12   Plot of n_{A0}/r_A versus conversion. The area under the curve corresponds with the volume of the reactor.

The irreversible gas-phase reaction,

$$A + B \rightarrow 2C \quad r_A = k_1 P_A P_B$$

is to be carried out in an adiabatic PFR. Using the data below calculate the reactor volume required for 80% conversion of the reactants (numerical integration is required).

Inlet mole fractions	10% A, 10% B, 80% inert
Feed temperature	300 K
Total molar feed rate	8.9×10^{-4} mol s^{-1}
Operating pressure	1 bar

Example 5.3

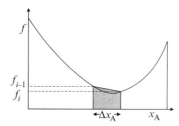

Fig. 5.13   Use of the trapezoidal rule to estimate reactor volume.

$$k_1 = 2 \times 10^6 \exp\left(-\frac{3000}{T}\right) \text{ mol s}^{-1} \text{ m}^{-3} \text{ bar}^{-2} \quad T \text{ in Kelvin}$$

Heat of reaction	-60 kJ (mol of A)$^{-1}$
Overall mean heat capacity	30 J mol^{-1} K^{-1}

Solution

$$V = -\int_{n_{A0}}^{n_A} \frac{dn_A}{r_A}$$

$$r_A = k_1 P_A P_B = k_1 P_A^2 = k_1(1-x_A)^2 P_{A0}^2$$

$$V = \int_0^{x_A} \frac{n_{A0}dx_A}{k_1(1-x_A)^2 P_{A0}^2}$$

$$= \frac{n_{A0}}{P_{A0}^2 k_1'} \int_0^{x_A} \frac{dx_A}{\exp\left(-\frac{3000}{T}\right)(1-x_A)^2}$$

Now we must find T in terms of x_A through the energy balance,

$$n_{A0}(-\Delta H_R)dx_A = n_{T0}\overline{c_P}dT$$

$$\frac{dT}{dx_A} = \frac{n_{A0}(-\Delta H_R)}{n_T \overline{c_P}} = \frac{(1)(60 \times 10^3)}{(10)(30)} \text{K} = 200 \text{ K}$$

$$T = T_0 + 200x_A$$

Now the function that we require to integrate can be evaluated as a function of conversion.

$$f(x_A, T) = \frac{n_{A0}}{P_{A0}^2 k_1'} \frac{1}{\exp\left(-\frac{3000}{T}\right)(1-x_A)^2}$$

$$= \frac{4.45 \times 10^{-9}}{\exp\left(-\frac{3000}{T}\right)(1-x_A)^2}$$

Trapezoidal rule, $V \approx \Delta x_A \left(\frac{f_0}{2} + f_1 + \cdots + f_7 + \frac{f_8}{2}\right)$

$$\approx 0.1(38.24 \times 10^{-5}) \approx 3.8 \times 10^{-5} \text{ m}^3$$

x_A	T	$f(x_A, T) \times 10^5$/m^3
0	300	9.80
0.1	320	6.48
0.2	340	4.72
0.3	360	3.78
0.4	380	3.32
0.5	400	3.22
0.6	420	3.52
0.7	440	4.52
0.8	460	7.56

5.2.3   Batch reactors

Again, for batch reactors we need to use an energy balance that is in differential form. Considering a differential change in temperature caused by a differential change in composition occurring in a differential element of time, dt,

$$N_{A0}dx_A(-\Delta H_R) = N_T\overline{c_P}dT + Qdt \tag{5.19}$$

Remembering that the material balance for a batch reactor (eqns 2.2 and 2.6) can be written as,

$$r_A Vdt = -dN_A = N_{A0}dx_A \tag{5.20}$$

and substituting eqn 5.20 into the energy balance (eqn 5.19) gives a relationship between temperature and time,

$$r_A V dt(-\Delta H_R) = N_T \overline{c_P} dT + Q dt$$

$$\frac{dT}{dt} = \frac{r_A V(-\Delta H_R) - Q}{N_T \overline{c_P}} \tag{5.21}$$

The material balance (eqn 5.20) relates conversion to time,

$$\frac{dx_A}{dt} = \frac{r_A V}{N_{A0}} \tag{5.22}$$

Compare eqns 5.21 and 5.22 to eqns 5.13 and 5.14. Equations 5.21 and 5.22 can be solved simultaneously in a manner analogous to the treatment for a PFR.

5.3 Steady-state multiplicity in CSTRs

Consider the first-order irreversible reaction,
$$A \rightarrow B$$
The material balance for a CSTR (eqn 3.10) with substitution for the temperature dependency of the reaction rate constant gives,

$$x_A = \frac{k\tau}{1 + k\tau} = \frac{1}{1 + \dfrac{1}{\tau k' \exp\left(-\dfrac{\Delta E}{RT}\right)}} \tag{5.23}$$

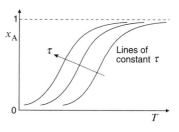

Fig. 5.14 Lines of constant residence time for a first-order reaction in a CSTR in conversion–temperature space.

Therefore, for a constant τ, the conversion in the reactor can be plotted as a function of temperature (see Fig. 5.14). At very low temperatures the conversion will asymptotically approach zero and at very high temperatures the conversion will asymptotically approach unity. Consequently, a family of sigmoidal curves are obtained, each one corresponding to a different value of τ. Obviously, for a reactor operating at a given temperature the conversion will be greater if the residence time, τ, is greater.

To find the operating point of the CSTR we need another relationship between conversion and temperature. This relationship is the energy balance, eqn 5.8,

$$n_{A0} x_A (-\Delta H_R) = n_T \overline{c_P}(T - T_0) + Q$$

In the case of cooling or heating coils,

$$n_{A0} x_A (-\Delta H_R) = n_T \overline{c_P}(T - T_0) + UA(T - T_j) \tag{5.24}$$

$$n_{A0} x_A (-\Delta H_R) = (n_T \overline{c_P} + UA)T - (n_T \overline{c_P} T_0 + UAT_j) \tag{5.25}$$

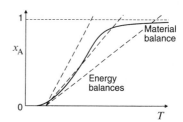

Fig. 5.15 Simultaneous plot of the material and energy balances for a CSTR with a single exothermic irreversible reaction.

Equation 5.25 is a straight-line relationship in conversion–temperature space. Figure 5.15 shows the material and energy balance plotted together for an exothermic reaction. Three different energy balances are shown, all with different gradients. This could be achieved by having, for example, different heat transfer areas (unless the reactor inlet temperature is adjusted this would also affect the intercept). Figure 5.16 shows the effect of altering the reactor inlet temperature or the reactor coolant temperature on the energy balance. In both cases (Figs 5.15 and 5.16) we see that up to three steady-state operating

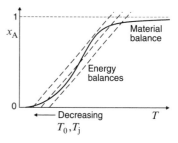

Fig. 5.16 Effect of altering reactor inlet temperature and coolant temperature on the energy balance.

[5] What we are doing here is slightly different from how we used the energy balance in Example 5.2. In that example, the conversion was specified and we chose an energy balance that would allow that conversion to be one of the possible steady-state operating points. We did not look to see if there were other possible steady states also available.

points (characterized by the material balance and energy balance being simultaneously satisfied) can be obtained, one at low conversion, one at intermediate conversion, and one at high conversion[5]. If only one steady-state operating point is available it can be at either low conversion or high conversion.

The low-conversion steady-state operating point is favoured by,

(a)   low rate constant
(b)   low residence time
(c)   small heat of reaction
(d)   small concentration of reactant
(e)   low values of inlet and coolant temperature
(f)   large heat transfer areas and heat transfer coefficient (for cooling)

The high-conversion operating point is favoured by the opposite conditions.

If multiple steady states are available how do we know at which one the reactor will be operating? We need to think about the stability of the operating points. Instead of thinking in terms of the material and energy balances let us think in terms of the heat generated by reaction and the heat removed by the cooling medium and the process stream.

The heat generated, Q_g, will be given by (using eqn 5.23),

$$Q_g = n_{A0}(-\Delta H_R)x_A = n_{A0}(-\Delta H_R)\frac{k\tau}{1 + k\tau} = \frac{n_{A0}(-\Delta H_R)}{1 + \dfrac{1}{\tau k' \exp\left(-\frac{\Delta E}{RT}\right)}} \quad (5.26)$$

and the heat removed by the cooling medium, Q, and the process stream, Q_r, will be given by,

$$Q_r + Q = n_T \overline{c_P}(T - T_0) + UA(T - T_j) \quad (5.27)$$

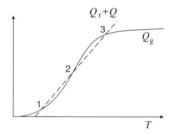

Fig. 5.17 Heat generation, Q_g, and heat removal, $Q_r + Q$, as a function of reactor temperature. The three steady-state operating points are shown.

Now if we plot the heat generation and heat removal lines against temperature (see Fig. 5.17), again we see that we can have the same three steady-state operating points.

Let us now look at the low-conversion operating point (operating point 1) in more detail. If the reactor is at that operating point and, due to some perturbation, it moves to a slightly higher temperature, then, at this slightly higher temperature, the rate of heat removal must be greater than the rate of heat generation. Hence the reactor temperature falls until the steady-state operating point 1 is again reached. If the initial perturbation resulted in the reactor temperature falling slightly, the heat generation would now be greater than the heat removal and the reactor temperature would increase until it again reached the steady-state operating point 1. Consequently, operating point 1 is stable.

Now let us look at the intermediate-conversion operating point in more detail. If the reactor temperature is perturbed to a higher level, the heat generation becomes greater than the heat removal and the reactor continues to heat up (it will eventually reach steady-state operating point 3 where it will remain). Conversely, if the reactor temperature drops slightly then the rate of heat removal will become greater than the rate of heat generation and the reactor will continue to cool until steady-state operating point 1 is reached. Therefore, the operating point at intermediate conversion is unstable.

By these arguments the operating point 3 will also be stable and we can write down the stability condition in a general form.

For stability, $\quad \dfrac{dQ_g}{dT} < \dfrac{d(Q_r + Q)}{dT}$ $\qquad$ (5.28)

i.e. the rate of heat generation must increase slower than the rate of heat removal with increasing temperature.

So now we know that our reactor at stable steady state will operate at either point 1 or point 3 (if we do not use external control to force operation at point 2). How do we ensure that the reactor is at the desired of the two operating points? If we start the reactor up, its temperature will increase until it reaches operating point 1 and it will stay there. However, in general we will want to operate at the high-conversion operating point—so how do we get there? If we preheat the inlet stream to the reactor we can ensure that the high-conversion operating point is the only possible operating point (see Fig. 5.18). (It could be that this operating point would be at too high a temperature for the reactor materials, in which case the reactant could be initially diluted.) Now once we are at the high-conversion operating point we can reduce the temperature of the inlet feed and the reactor will remain at the high-conversion operating point. What we have done is used preheating to cause 'ignition'.

Up until now we have only considered a single exothermic reaction. Let us now look at the heat generation and removal lines (or material and energy balances) for some other cases of reaction.

Consider a series reaction (with both steps being exothermic), e.g.

$$A \rightarrow B \rightarrow C$$

If the first reaction goes almost to completion before the second reaction starts appreciably, then the corresponding curves are as shown in Fig. 5.19(a). It can be seen that up to five steady-state operating points could be obtained. Three would be stable and two would be unstable.

For an endothermic irreversible reaction the heat generation is negative so heat must be supplied for steady operation; see Fig. 5.19(b). This means that only one operating point can ever be obtained and it must always be stable.

For a reversible exothermic reaction see Fig. 5.19(c). The heat generation will increase to some temperature where it will go through a maximum and then start decreasing as equilibrium is approached at higher temperatures. Remember that heat generation is proportional to rate of reaction and that rate of reaction will go through a maximum as the temperature is varied. Therefore, if we design a CSTR to operate at this maximum in reaction rate it must satisfy the following conditions,

$$\frac{dr_A}{dT} = 0; \quad \frac{dQ_g}{dT} = 0 \quad \text{and} \quad \frac{d(Q_r + Q)}{dT} > 0$$

and is therefore stable.

The exothermic gas-phase reaction,

$$A \Leftrightarrow B + C; \quad r_A = k_1 P_A - k_{-1} P_B P_C$$

is to be carried out in a CSTR operating at a pressure of 50 bar.

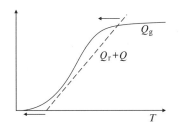

Fig. 5.18 Schematic showing how preheating the feed stream will lead to only one steady-state operating point. Allowing the temperature of the feed stream to then decrease ensures that the reactor will operate at the high-conversion operating point.

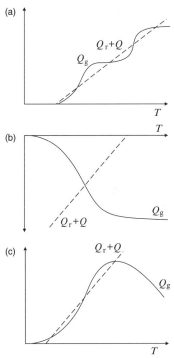

Fig. 5.19 Heat generation, Q_g, and heat removal, $Q_r + Q$, as a function of temperature for: (a) two exothermic reactions in series; (b) an endothermic reaction; (c) a reversible exothermic reaction.

Example 5.4

(a) Calculate the minimum reactor volume if the reactor operates at a 22% conversion of A (feed is pure A fed at 5 mol s^{-1}).

(b) Heat is removed from the reactor via cooling coils. What heat transfer area is required for steady operation under the above conditions if the overall heat transfer coefficient is 10 W m^{-2} K^{-1}?

(c) Comment on the stability of the operating point.

$$k_1 = 0.435 \exp\left(\frac{-20\,000}{RT}\right) \text{ mol s}^{-1} \text{ m}^{-3} \text{ bar}^{-1}$$

$$k_{-1} = 147 \exp\left(\frac{-60\,000}{RT}\right) \text{ mol s}^{-1} \text{ m}^{-3} \text{ bar}^{-2}$$

RT is in J mol^{-1}; $R = 8.314$ J mol^{-1} K^{-1}

Feed temperature	350 K
Coolant temperature	400 K
Heat of reaction	–40 000 J (mol of A)$^{-1}$
Mean heat capacity of A, B, C	30, 20, 10 J mol^{-1} K^{-1} respectively

Solution (a) In Example 5.1, we have already shown that the maximum rate of reaction,

$$r_A = 0.183 \text{ mol s}^{-1} \text{ m}^{-3}$$

As 1.1 mol s^{-1} of A are converted, the corresponding minimum volume is,
$$V = 5.94 \text{ m}^3$$

(b) Heat released

$$n_{A0} x_A (-\Delta H_R) = (1.1)(40\,000) = 44 \text{ kW}$$

To heat up reactants,

$$\overline{c_{PA}} n_{A0} \Delta T = (30)(5)(264.8) = 39.7 \text{ kW}$$

Must remove 4.28 kW by heat transfer,

$$4.28 \text{ kW} = UA\Delta T$$
$$A = 1.99 \text{ m}^2$$

(c) The operating point must be stable as

$$\frac{dQ_g}{dT} = 0 \quad \text{and} \quad \frac{d(Q_r + Q)}{dT} > 0$$

5.4   Multistage adiabatic PFR

For large-scale industrial reactors, heat supply or removal through the reactor wall can be extremely difficult and operation must effectively be adiabatic. However, we know that for an exothermic reversible reaction there is, for any given conversion, a temperature that maximizes the reaction rate (Fig. 5.4 and the solution of eqn 5.1). This means that for a PFR there exists an optimum temperature profile or line of maximum reaction rate (if we have the aim of minimizing the total reactor volume) and it is important to try to approach this path. Two different methods for achieving this, interstage cooling and cold-shot cooling, will be discussed here.

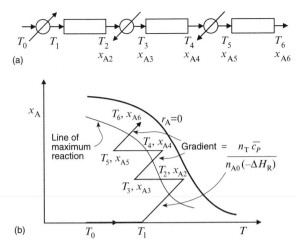

Fig. 5.20 (a) Process diagram for interstage cooling. (b) Operating lines for interstage cooling in conversion–temperature space.

5.4.1 Interstage cooling

We preheat our process stream so that the initial reaction rate is not too low. The stream is allowed to react until the temperature becomes high enough that the rate begins to decrease significantly as equilibrium is approached (this point where the reactor should be terminated to give a minimum overall volume can be rigorously determined—however, such a rigorous approach is beyond the scope of this text). The reactor is then 'terminated' (i.e. the first stage of the PFR is designed to be of the appropriate size) and the exiting stream is cooled in a heat exchanger to take it away from equilibrium. The stream is then fed to the second reactor stage in the series and once more allowed to react, heat up, and approach equilibrium. Once again, when the reaction rate has fallen sufficiently the process stream is passed on to a heat exchanger for cooling before the next reaction stage. Figure 5.20(a) shows a simple process diagram and the behaviour, in terms of a plot of conversion against temperature, is shown in Fig. 5.20(b). For adiabatic reactors, all of the reaction stages will have the same gradient (eqn 5.16). The lines representing the cooling stages have a gradient of zero as the conversion does not change ($x_{A2} = x_{A3}$, $x_{A4} = x_{A5}$). In Fig. 5.20(b) it can be seen that this kind of strategy has the result of keeping the operating conditions close to the line of maximum reaction rate.

However, heat exchangers can be expensive or alternatively, if we are considering a high-pressure process, the volume they occupy in a pressure vessel can have a very high capital cost associated with it. These factors must be considered in deciding whether interstage cooling should be used versus the alternative approach (which does not rely on the use of heat exchangers) of cold-shot cooling.

5.4.2 Cold-shot cooling

The principle behind cold-shot cooling is that only a fraction of the process stream is fed to the first reactor stage. When this fraction exits the reactor it is cooled by using cold fresh feed rather than a heat exchanger (see Fig. 5.21(a)).

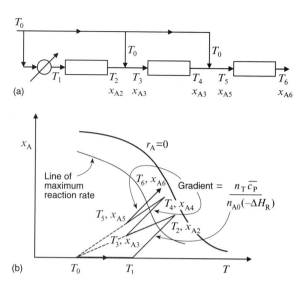

Fig. 5.21 (a) Process diagram for cold-shot cooling. (b) Operating lines for cold-shot cooling in conversion–temperature space.

This serves to eliminate the need for a heat exchanger but could be viewed as less efficient than interstage cooling because in mixing with fresh feed the conversion effectively drops. In practice, cold-shot reactors are used for high-pressure processes such as ammonia synthesis. This reduces the total volume of the pressure vessel as interstage heat exchangers are no longer needed. Let us now look at what is happening in terms of temperature versus conversion (see Fig. 5.21(b)). We preheat a fraction of the process stream from T_0 to T_1 and then allow it to react. When it exits the first reactor stage it is at a temperature T_2 and a conversion x_{A2}. This stream is now mixed with the cold fresh stream at temperature T_0 and with zero conversion. The resulting mixed stream, (x_{A3}, T_3), must lie on a straight line between the two streams that are mixed (i.e. between $(0, T_0)$ and (x_{A2}, T_2)), in conversion–temperature space. The position of the mixed stream on the line depends upon the relative contributions from the two streams. The larger the relative fraction of the original fresh stream the closer the conditions of the mixed stream will be to it. This stream is then fed into the second reaction stage. Cooling is then again achieved in the same manner. Again, the gradients of the reactor operating lines will all be similar in the case of adiabatic operation.

Example 5.5   An exothermic, reversible, gas-phase reaction is to be performed in a multistage adiabatic PFR,

$$A \Leftrightarrow B$$

$$K = 25 \times 10^{-3} \exp\left(\frac{-\Delta H_R}{RT}\right)$$

$$\Delta H_R = -20 \text{ kJ (mol of A)}^{-1}$$

$$\overline{c_P} = 140 \text{ J mol}^{-1} \text{ K}^{-1}$$

$$R = 8.314 \text{ J mol}^{-1} \text{ K}^{-1}$$

The feed is pure A. The temperature of the feed and coolant streams is 290 K.

Calculate the maximum possible conversion of A using: (a) interstage cooling; (b) cold-shot cooling.

Solution

Equilibrium condition,
$$K = \frac{n_B^*}{n_A^*} = \frac{x_A^*}{1 - x_A^*}$$

$$x_A^* = \frac{K}{1 + K} = \frac{1}{1 + 400 \exp\left(-\dfrac{2406}{T}\right)}$$

(a)   With interstage cooling the maximum conversion achievable is when the process stream is at equilibrium at 290 K (to obtain this a large reaction volume and plenty of cooling would be required),

$$K = 10.0, \quad x_A = 0.91$$

(b)   For cold-shot cooling the maximum conversion can be reached using one reactor stage with inlet temperature of 290 K. This reactor stage, if of a volume approaching infinity, would give a conversion that would approach equilibrium. No heat is actually removed from the process stream with cold-shot cooling; therefore, we can never do better than is possible with the cold stream fed to one reactor stage as far as the greatest conversion is concerned (of course, we can do much better using a series of reactor stages if we are concerned about the total reactor volume). Alternatively, we can never get past the operating line for a single reactor stage with the cold fresh feed as its inlet stream as the operating line is coincidental with the mixing line.

The operating line is,

$$n_{A0} x_A (-\Delta H_R) = n_T \overline{c_P} (T - T_0)$$

$$x_A = \frac{\overline{c_P}(T - T_0)}{-\Delta H_R} = 7 \times 10^{-3}(T - 290)$$

Operating line, $x_A' = 7 \times 10^{-3}(T - 290)$

Equilibrium line, $x_A'' = \dfrac{1}{1 + 400 \exp\left(-\dfrac{2406}{T}\right)}$

Solve by trial and error, $x_A = 0.60$

T	X_A'	X_A''
290	0	0.91
400	0.77	0.51
350	0.42	0.71
375	0.60	0.60

5.5   Problems

5.1   A CSTR is used for carrying out the liquid-phase reaction

$$A + B \rightarrow \text{Products} \quad r_A = k C_A C_B$$

The feed concentration of both A and B is 3 mol litre^{-1}. The volumetric flow rate is 1 litre s^{-1}.

(a)   Calculate the operating temperature required for a conversion of 60%.

(b)   How much heat must be added to the reactor to maintain the system at steady state if the feed temperature is 290 K?

$V = 4$ litres
$k = 10^{14} \, e^{-\Delta E/RT}$ litre mol^{-1} s^{-1}
$\Delta H_R = 60$ kJ per mol of A reacted
$\Delta E = 100$ kJ mol^{-1}
$c_P = 4.2$ kJ litre^{-1} K^{-1}

5.2 The reversible gas-phase reaction

$$A \Leftrightarrow B \quad r_A = k_1 \, P_A - k_{-1} \, P_B$$

is to be carried out in a CSTR. The feed is 10% A, 90% inert, and a conversion of 62% is required.
Calculate: (a) the minimum necessary reactor volume; (b) the amount of heat to be supplied or removed for steady operation.

$n_T = 2.1$ mol s^{-1}
Feed temperature, $T_0 = 573$ K
Pressure, $P = 1$ bar
$k_1 = 10^3 \exp(-\Delta E/RT)$ mol m^{-3} s^{-1}
$\qquad$ bar^{-1}
$k_{-1} = 10^6 \exp(-\Delta E_{-1}/RT)$ mol m^{-3}
$\qquad\qquad$ s^{-1} bar^{-1}
$\Delta E_1 = 41\,570$ J mol^{-1};
$\Delta E_{-1} = 83140$ J mol^{-1}
$\overline{C_{PA}} = \overline{C_{PB}} = \overline{C_{Pinert}} = 42$ J mol^{-1}
$\qquad\qquad\qquad\qquad$ K^{-1}

5.3 It is desired to carry out the same reaction as in Problem 5.2, again with 62% conversion. This time a CSTR with a volume of 5 m^3 is available and is to be used in series with another CSTR, which is to be designed.
(a) What is the minimum volume required of this other CSTR if the feed stream is fed to it initially (see Fig. 5.22)?
(b) What are the feed temperatures for the two reactors if they are operated adiabatically?
(c) What would be the minimum volume required of this other CSTR if the feed stream were fed to the 5 m^3 CSTR first? (Iterative procedure required.)

5.4 The gas-phase reaction

$$A \rightarrow B \quad r_A = kP_A$$

is to be carried out in an adiabatic PFR. The feed consists of 50% A and 50% inert at a total flow rate of 6.3 mol s^{-1} and a

temperature of 300 K. A conversion of 80% is required.
Calculate (numerical integration required) the reactor volume required.

$P = 1$ bar
$k = 28.1 \, e^{-\Delta E/RT}$ mol s^{-1} m^{-3} bar
$\Delta H_R = -50$ kJ mol^{-1}
$\Delta E/R = 1000$ K
$\overline{c_P} = 100$ J mol^{-1} K^{-1}

5.5 A gas-phase reaction

$$A \rightarrow 2B \quad r_A = kP_A$$
$$k = 4.9 \exp\left(-\frac{20\,000}{RT}\right) \text{ (mol of A)}$$
$$\text{m}^{-3}\text{s}^{-1} \text{ bar}^{-1}$$
$$RT \text{ in J mol}^{-1}$$

is performed in a batch reactor. Initially, the reactor and its contents of pure A are at 300 K. The reaction is then initiated and the reactor temperature is linearly increased with time at a rate of 0.2 K s^{-1} to a final temperature of 540 K (giving a total residence time of 1200 s). Ideal gas behaviour may be assumed. Calculate the final conversion of A (numerical integration is required).

5.6 The gas-phase reaction

$$A \rightarrow B + C \quad r_A = kC_A$$

is to be carried out in a CSTR of volume 0.02 m^3. Heat is to be removed from the reactor by cooling coils carrying water at 373.2 K and discharging saturated steam at 373.2 K. The feed consists of pure A at a temperature of 325 K and at a pressure of 1 bar and flow rate 10^{-3} mol s^{-1}.
Calculate (an iterative procedure is required) the operating temperature and conversion of the steady-state operating points.

$UA = 5 \times 10^{-2}$ W K^{-1}
$$k = 10^{11} \exp\left(-\frac{18\,000}{T}\right) \text{ s}^{-1}$$
T is in Kelvin
$\Delta H_R = -70$ kJ (mol of A)$^{-1}$
$\overline{c_{PA}} = 120$ J mol^{-1} K^{-1}
$\overline{c_{PB}} = 80$ J mol^{-1} K^{-1}
$\overline{c_{PC}} = 40$ J mol^{-1} K^{-1}
$R = 8.314$ J mol^{-1} K^{-1}

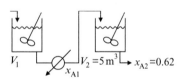

V_1 $\quad$ $V_2 = 5\,\text{m}^3$ $\quad x_{A2} = 0.62$
$\quad x_{A1}$

Fig. 5.22 Schematic representation of Problem 5.3(a).

6 Non-ideal reactors

In practice, plug flow and perfect mixing are never achieved. The behaviour of all reactors is somewhere between these two extremes of no mixing and ideal mixing. This is because of effects such as stagnant regions and 'short circuiting' (see Fig. 6.1). In a tubular reactor these stagnant regions will have a longer residence time than the rest of the process stream and mixing will occur within the stagnant regions. In stirred tank reactors some elements might 'short circuit' or bypass the well-mixed bulk of the reactor. This will result in a larger fraction of the process stream having short residence times and there will be incomplete mixing of all the elements in the reactor.

We need an approach to understand the behaviour of real reactors. The approach we use is built upon the concept of residence time distributions (RTDs). We can model a continuous reactor as a collection of small elements; each one of these elements will take a particular path through the reactor and have an associated residence time (see Fig. 6.2). The RTD is a distribution plot of the fraction of these elements exiting the reactor with different residence times.

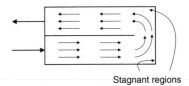

Stagnant regions

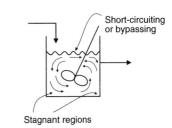

Short-circuiting or bypassing

Stagnant regions

Fig. 6.1 Short-circuiting, by-passing, and stagnant regions in reactors.

6.1 Residence time distributions (RTDs)

Residence time distributions, or RTDs, can be obtained by injecting a tracer into the process stream at the reactor inlet. For instance, if we have a normally colourless process stream, we could inject an instantaneous pulse of red dye at the reactor inlet. Effectively, we have labelled a representative group of elements entering the reactor at the time of the pulse. We then monitor the concentration of the dye in the outlet. The concentration of dye in the outlet at any particular time will be proportional to the fraction of labelled elements leaving the reactor at that particular time.

6.1.1 Ideal CSTR

For an ideal CSTR, after injection of dye, we would instantaneously see some dye in the outlet stream because perfect mixing would mean that all of the dye immediately mixed with the liquid in the reactor; the concentration of dye in the outlet stream from the reactor would then drop exponentially with time as the dye continuously leaves the reactor while normal feed, with no dye present, enters the reactor (Fig. 6.3(a)).

Mathematically, let us consider N moles of tracer instantaneously introduced into the reactor inlet at time $t = 0$. Instantaneous mixing means that the initial concentration in the reactor is uniform,

$$\frac{N}{V} = C_0$$

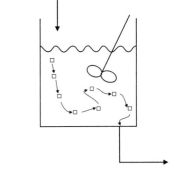

Fig. 6.2 Modelling of a continuous reactor by considering it to behave as a collection of differential batch reactors with different residence times.

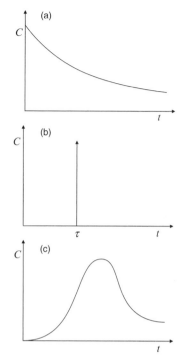

Fig. 6.3 Concentration versus time for tracer injected into: (a) an ideal CSTR; (b) an ideal PFR; (c) a real reactor.

The unsteady material balance says that the rate of accumulation of dye must equal minus the rate of out-flow of the dye (there is no in-flow and the dye is not involved in any reaction),

$$V\frac{dC}{dt} = -v_T C \tag{6.1}$$

Equation 6.1 is easily integrated for constant volumetric flow rate to give,

$$C(t) = C_0 \exp\left(-\frac{v_T t}{V}\right) = C_0 \exp\left(-\frac{t}{\tau}\right) \tag{6.2}$$

If we want to compare different experiments (which could be conducted with different amounts of injected dye) it is important to normalize this distribution. The normalized form is known as the residence time distribution (RTD) and is denoted by the variable, E.

$$E(t) = \frac{C(t)}{\int_0^\infty C(t)dt} \tag{6.3}$$

But
$$\int_0^\infty C(t)dt = \int_0^\infty \frac{n(t)}{v_T}dt$$

where $n(t)$ is the molar flow of tracer in the outlet of the stream.

However
$$\int_0^\infty n(t)dt = N$$

Therefore
$$\int_0^\infty C(t)dt = \frac{N}{v_T} = \frac{VC_0}{v_T} = \tau C_0 \tag{6.4}$$

and substituting eqns 6.4 and 6.2 into eqn 6.3,

$$E(t) = \frac{1}{\tau} \exp\left(-\frac{t}{\tau}\right) \tag{6.5}$$

The RTD decays exponentially and because of the normalization the area under the curve is equal to unity.

6.1.2　Ideal PFR

For an ideal PFR, if an instantaneous pulse of tracer were injected, we would see no dye until the residence time of the reactor had passed and then all of the dye would come out in a single pulse (see Fig. 6.3(b)). The RTD is simply given by the delta function,

$$E(t) = \delta(t - \tau) \tag{6.6}$$

where $\delta(t - \tau) = 0$, for $t \neq \tau$

$\quad\quad \delta(t - \tau) \neq 0$, for $t = \tau$

with $\int_0^\infty \delta(t - \tau)dt = 1$

This corresponds to a pulse of tracer (which, because it is introduced instantaneously and it does not mix with any other fluid elements in the reactor, is of infinite concentration) and the area under the curve is unity.

6.1.3   Real reactor

In the case of a real reactor the situation is somewhere between these two extremes (Fig. 6.3(c)). Because there is a degree of mixing in the reactor we do not see a sharp peak, as in the case of a PFR; instead, it is more disperse. Mixing is not instantaneous, as in a CSTR, so we do not see an instantaneous value for the outlet concentration of the dye at time $t = 0$.

6.2   Calculation of the mean residence time

The RTD can be used to calculate the mean residence time of any reactor. Returning to the discussion about labelling the elements with a pulse of red dye in the inlet, the fraction of dye that leaves the reactor in any time interval is simply proportional to the average concentration of dye in the outlet stream during the interval multiplied by the duration of the interval. Therefore, the fraction of the total stream that has a residence time between t and $t + dt$ is simply $E(t)dt$ (see Fig. 6.4). To calculate the mean residence time, we take a weighted average over all residence times, so that the mean residence time is simply the individual residence time for each element multiplied by the fraction of the dye in that element, summed over all elements. Expressing this for infinitely small elements in integral form,

$$\bar{t} = \int_0^\infty tE(t)dt \tag{6.7}$$

If we take the case of a CSTR on substitution of eqn 6.5 we get,

$$\bar{t}_{CSTR} = \int_0^\infty \frac{t}{\tau} \exp\left(-\frac{t}{\tau}\right)dt$$

Integrating by parts,

$$\bar{t}_{CSTR} = \left[-t\exp\left(-\frac{t}{\tau}\right)\right]_0^\infty + \int_0^\infty \exp\left(-\frac{t}{\tau}\right)dt$$

$$= \left[-t\exp\left(-\frac{t}{\tau}\right) - \tau\exp\left(-\frac{t}{\tau}\right)\right]_0^\infty$$

$$= [0 - (-\tau)] = \tau = \frac{V}{v_T}$$

which is exactly the result we would expect for the mean residence time of the CSTR.

Likewise, we can use the RTD of a PFR, substituting eqn 6.6 into eqn 6.7,

$$\bar{t}_{PFR} = \int_0^\infty t\delta(t - \tau)dt = \tau = \frac{V}{v_T}$$

again as expected.

Given the reactor residence time distribution (RTD) shown in Table 6.1, evaluate the mean residence time.

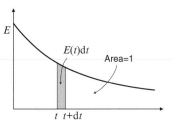

Fig. 6.4   Residence time distribution for an ideal CSTR. The fraction of the total stream that has a residence time between t and $t + dt$ is $E(t)dt$.

Table 6.1

t/s	E/s^{-1}
0	0.00
2	0.04
4	0.15
6	0.15
8	0.10
10	0.05
12	0.01
14	0.00

Example 6.1

Solution   Equation 6.7, $\bar{t} = \int_0^\infty tE(t)\mathrm{d}t$

However, we have been supplied with discrete data and therefore we must use a summation rather than an integral.

$$\bar{t} \approx \Delta t \sum tE$$

Evaluating tE, as in Table 6.2, gives $\bar{t} = 6$ s

Table 6.2

t/s	tE
0	0.00
2	0.08
4	0.60
6	0.90
8	0.80
10	0.50
12	0.12
14	0.00

6.3   Calculation of conversion from RTD

We have already said that we can model a continuous reactor as a collection of small elements. The concentration of reactant in each of these elements will be a function of how long the element has been in the reactor and, from the RTD, we know how long each element will spend on average in the reactor. Therefore, we can calculate the mean outlet concentration of reactant,

$$\overline{C_A} = \int_0^\infty C_A(t)E(t)\mathrm{d}t \tag{6.8}$$

We may treat each of these elements as batch reactors and, if they are small enough, we can consider them to be well mixed internally. For a first-order, irreversible reaction the design equation for a well-mixed batch reactor is,

$$\frac{\mathrm{d}C_A}{\mathrm{d}t} = -r_A = -kC_A$$

Therefore, for any element,

$$C_A = C_{A0}\exp(-kt)$$

and substituting into eqn 6.8,

$$\overline{C_A} = C_{A0}\int_0^\infty \exp(-kt)E(t)\mathrm{d}t \tag{6.9}$$

Table 6.3

t/s	E/s^{-1}
0	0.00
2	0.04
4	0.15
6	0.15
8	0.10
10	0.05
12	0.01
14	0.00

Evaluation of this integral for a PFR is trivial and so we will demonstrate the case of a CSTR. We substitute eqn 6.5 into eqn 6.9,

$$\overline{C_{A,CTSR}} = C_{A0}\int_0^\infty \exp(-kt)\frac{1}{\tau}\exp\left(-\frac{t}{\tau}\right)\mathrm{d}t$$

$$= C_{A0}\int_0^\infty \frac{1}{\tau}\exp\left(-kt-\frac{t}{\tau}\right)\mathrm{d}t$$

$$= C_{A0}\left[-\frac{1}{1+k\tau}\exp\left(-kt-\frac{t}{\tau}\right)\right]_0^\infty$$

$$= \frac{C_{A0}}{1+k\tau}$$

which is the expression that would be expected for a CSTR.

Example 6.2   A first-order, liquid-phase reaction,

$$A \rightarrow B, \quad r_A = kC_A, \quad k = 0.307\ \mathrm{s}^{-1}$$

is performed in a non-ideal reactor with the reactor residence time distribution (RTD) shown in Table 6.3.

(a) What is the conversion of A?
(b) Compare the conversion to that of a CSTR and a PFR of the same mean residence time (6 s).

(a) For a first-order irreversible reaction, eqn 6.9 applies,

$$\frac{\overline{C_A}}{C_{A0}} = \int_0^\infty \exp(-kt)E(t)dt$$

From Tables 6.3 and 6.4,

$$\frac{\overline{C_A}}{C_{A0}} \approx \Delta T \sum \exp(-kt)E = 2(0.106) = 0.21$$

$$x_A = 1 - \frac{\overline{C_A}}{C_{A0}} \approx 0.79$$

(b) $x_{A,PFR} = 1 - \exp(-k\tau) = 0.84$

$$x_{A,CSTR} = \frac{k\tau}{1 + k\tau} = 0.65$$

We can see that the conversion estimated for the real reactor lies between the two extremes of the CSTR and the PFR.

Solution

Table 6.4

t/s	$e^{-kt}E/s^{-1}$
0	0.00
2	0.022
4	0.045
6	0.026
8	0.009
10	0.003
12	0.001
14	0.000

Solutions

Solutions to Chapter 3

3.1 $A \to 2B$ $r_A = kP_A = k\dfrac{N_A}{N_T}P$

(a) $N_A = N_{A0} - N_{A0}x_A$

$N_B = 2N_{A0}x_A$

$N_I = N_{I0}$

$N_T = N_{T0} + N_{A0}x_A$

$\dfrac{P_T}{P_0} = \dfrac{N_{T0} + N_{A0}x_A}{N_{T0}} = 1 + \dfrac{x_A}{2} = 1.38, x_A = 0.76$

(b) $r_A = kP_A = k\dfrac{N_A}{N_T}P; \quad PV = N_T RT$

$r_A = kN_A\dfrac{RT}{V}; r_A V = kRTN_A$

(c) $r_A = -\dfrac{1}{V}\dfrac{dN_A}{dt}; t = -\displaystyle\int_{N_{A0}}^{N_A}\dfrac{dN_A}{r_A V}$

$t = -\dfrac{1}{kRT}[\ln N_A]_{N_{A0}}^{N_A} = \dfrac{1}{kRT}\ln\dfrac{N_{A0}}{N_A}$

$t = \dfrac{1}{kRT}\ln\left(\dfrac{1}{1-x_A}\right); k = \dfrac{1}{tRT}\ln\left(\dfrac{1}{1-x_A}\right)$

$k = \dfrac{1}{(170)(8.314)(500)}\ln(4.167) = 2.02 \times 10^{-6} \text{ mol Pa}^{-1} \text{ m}^{-3} \text{ s}^{-1}$

$k = 0.202 \text{ mol bar}^{-1} \text{ m}^{-3} \text{ s}^{-1}$

3.2 $t = \displaystyle\int_{N_{A0}}^{N_A}\dfrac{1}{V}\dfrac{dN_A}{r_A}$

$2A \to B; r_A = kC_A^2 = k\dfrac{N_A^2}{V^2}$

$t = \displaystyle\int_{N_{A0}}^{N_A}\dfrac{1}{V}\dfrac{dN_A}{kN_A^2}V^2$

Constant volume, V_0

$$t_v = -\frac{V_0}{k}\left[-\frac{1}{N_A} + \frac{1}{N_{A0}}\right]$$

$$= \frac{V_0}{k}\left(\frac{1}{N_A} - \frac{1}{N_{A0}}\right)$$

$$= \frac{V_0}{k}\frac{1}{N_{A0}}\left(\frac{1}{1-x_A} - 1\right)$$

$$= \frac{V_0}{kN_{A0}}\frac{x_A}{1-x_A}$$

Constant pressure, P_0 (initially reactor is of volume V_0)

$$N_A = N_{A0} - N_{A0}x_A$$

$$N_B = N_B + \frac{1}{2}N_{A0}x_A$$

$$N_T = N_{A0} - \frac{1}{2}N_{A0}x_A$$

$$P_0V_0 = N_{A0}RT; \; P_0V = N_TRT$$

$$t_p = -\int_{N_{A0}}^{N_A}\frac{N_TRT}{P_0}\frac{dN_A}{kN_A^2}$$

$$= -\frac{RT}{P_0}\frac{1}{k}\int_0^{x_A}\frac{(1-\frac{1}{2}x_A)}{N_{A0}(1-x_A)^2}(-N_{A0})dx_A$$

$$= -\frac{RT}{P_0}\frac{1}{k}\int_0^{x_A}\left\{\frac{\frac{1}{2}}{1-x_A} + \frac{\frac{1}{2}}{(1-x_A)^2}\right\}dx_A$$

$$= -\frac{RT}{2P_0}\frac{1}{k}\left[-\ln(1-x_A) + \frac{1}{(1-x_A)}\right]_0^{x_A}$$

$$= \frac{V_0}{2kN_{A0}}\left\{\frac{x_A}{1-x_A} - \ln(1-x_A)\right\}$$

$$x_A = 0.9; \; t_v = 9V_0/kN_{A0}; \; t_p = 5.65V_0/kN_{A0}; \; t_v/t_p = 1.59$$

As the reaction proceeds the volume of the constant pressure reactor decreases, hence the concentration and therefore the rate are higher.

3.3 $2A \Leftrightarrow B$

$$r_A = k_1P_A^2 - k_{-1}P_B$$

$$\frac{n_{A0}}{n_{T0}} = \frac{2}{3}; \frac{n_{B0}}{n_{T0}} = \frac{1}{3}$$

$$n_A = n_{A0} - n_{A0}x_A, \; n_B = n_{B0} + \frac{1}{2}n_{A0}x_A, \; n_T = n_{T0} - \frac{1}{2}n_{A0}x_A$$
$$P = 1 \text{ bar}, \; x_A = 0.25$$

(a) $P_A = \dfrac{n_{A0}(1-x_A)}{n_{A0}(\frac{3}{2} - \frac{1}{2}x_A)}P = 0.545 \text{ bar}$

$$P_B = \frac{n_{A0}(\frac{1}{2} + \frac{1}{2}x_A)}{n_{A0}(\frac{3}{2} - \frac{1}{2}x_A)} P = 0.454 \text{ bar}$$

(b) $r_A = k_1 P_A^2 - k_{-1} P_B = (0.05)(0.545)^2 - (0.025)(0.454) \text{ mol m}^{-3} \text{ s}^{-1}$

$\qquad = 0.0035 \text{ mol m}^{-3} \text{ s}^{-1}$

(c) $n_{T0} = \dfrac{P v_{T0}}{RT}$

$$n_{A0} = \frac{2}{3} \frac{P v_{T0}}{RT} = \frac{2(10^5)(10^{-3})}{3(8.314)(600)} \text{ mol s}^{-1} = 0.0134 \text{ mol s}^{-1}$$

(d) $V = \dfrac{n_{A0} x_A}{r_A} = \dfrac{(0.0134)(0.25)}{(0.0035)} \text{ m}^3 = 0.96 \text{ m}^3$

3.4   Liquid phase, v_T is constant.

$$V = -\int_{n_{A0}}^{n_A} \frac{dn_A}{r_A}$$

$n_A = n_{A0} - n_{A0} x_A, \; n_B = n_{B0} - n_{A0} x_A$

$n_{A0} = n_{B0}$

$$V = -\int_0^{x_A} \frac{-n_{A0} dx_A}{k \frac{n_{A0}^2 (1-x_A)^2}{v_T^2}}$$

$$= \frac{v_T^2}{n_{A0} k} \int_0^{x_A} \frac{dx_A}{(1-x_A)^2}$$

$$= \frac{v_T^2}{n_{A0} k} \left(\frac{1}{1-x_A} - 1\right) = \frac{v_T}{C_{A0} k} \left(\frac{1}{1-x_A} - 1\right)$$

$n_{A0} = v_T C_{A0}$

$$V = \frac{(10^{-5})}{10^3 (10^{-6})} \cdot \left(\frac{1}{1-0.5} - 1\right) = 0.01 \text{ m}^3$$

3.5 $A + B \rightarrow C$

$$V = -\int_{n_{A0}}^{n_A} \frac{dn_A}{r_A} = -\int_{n_{A0}}^{n_A} \frac{dn_A}{k_P P_B}$$

$n_A = n_{A0} - n_{A0} x_A$

$n_B = n_{B0} - n_{A0} x_A$

$n_C = n_{A0} x_A$

$n_T = n_{T0} - n_{A0} x_A$

$n_{A0} = 2 n_{B0}$

$$n_{T0} = 3n_{B0} = \frac{3}{2}n_{A0}$$

$$P_B = \frac{n_B}{n_T} \cdot P = \frac{n_{A0}(\frac{1}{2} - x_A)}{n_{T0} - n_{A0}x_A}P = \frac{\frac{1}{2} - x_A}{\frac{3}{2} - x_A} \cdot P$$

$$V = -\int_0^{x_A} \frac{-n_{A0}(\frac{3}{2} - x_A)dx_A}{kP(\frac{1}{2} - x_A)}$$

$$= \frac{n_{A0}}{kP}\int_0^{x_A}\left(1 + \frac{1}{\frac{1}{2} - x_A}\right)dx_A$$

$$= \frac{n_{A0}}{kP}[x_A - \ln(\frac{1}{2} - x_A)]$$

$$= \frac{80}{(2.3 \times 10^3)(20)}[0.45 - \ln(0.05)]$$

$$= 6.0 \times 10^{-3} \text{ m}^3$$

3.6 $$V = -\int_{n_{Ai}}^{n_{Ae}} \frac{dn_A}{r_A}$$

$n_A = n_T y_A$, where y_A is the mole fraction of A

$$V = -\int_{y_{Ai}}^{y_{Ae}} \frac{n_T dy_A}{P(k_1 y_A - k_{-1}(1 - y_A))}$$

$$= -\int_{y_{Ai}}^{y_{Ae}} \frac{n_T dy_A}{P(k_1 + k_{-1})\left(y_A - \frac{k_{-1}}{(k_1 + k_{-1})}\right)}$$

$$\frac{k_{-1}}{(k_1 + k_{-1})} = 0.4$$

$$V = -\frac{n_T}{P(k_1 + k_{-1})} \ln \frac{y_{Ai} - 0.4}{y_{Ae} - 0.4}$$

Therefore, $$y_{Ae} = \frac{y_{Ai} - 0.4}{\varphi} + 0.4$$

where, $$\varphi = \exp\left(\frac{PV(k_1 + k_{-1})}{n_T}\right)$$

and we have a relationship between y_{Ae} and y_{Ai}.

(a) $n_T = n_{A0}$; $y_{Ai} = 1$

$$\frac{PV(k_1 + k_{-1})}{n_T} = 2$$

$$y_{Ae} = \frac{1 - 0.4}{e^2} + 0.4 = 0.481$$

$n_{A0}x_{Ae} = n_T(y_{Ai} - y_{Ae})$ (amount of A reacted)

$$x_{Ae} = 1 - y_{Ae} = 0.519$$

(b) $n_{Ti} = n_{Te} = n_T$

The flow in the reactor is no longer equal to n_{A0},

$$n_{A0} = n_T(1 - \alpha)$$

and rewriting the relationship between y_{Ae} and y_{Ai},

$$y_{Ae} = \frac{y_{Ai} - 0.4}{\varphi} + 0.4$$

where $\varphi = \exp\left(\dfrac{PV(k_1 + k_{-1})(1 - \alpha)}{n_{A0}}\right)$

The material balance at the mixing point,

$$n_{Ai} = n_{A0} + \alpha n_{Ae}$$

In terms of mole fractions,

$$y_{Ai} = \frac{n_{A0} + \alpha n_{Ae}}{n_T} = (1 - \alpha) + \alpha y_{Ae}$$

Eliminating y_{Ai},

$$y_{Ae} = \frac{(1 - \alpha) + \alpha y_{Ae} - 0.4}{\varphi} + 0.4$$

$$= \frac{0.4(\varphi - 1) + (1 - \alpha)}{\varphi - \alpha}$$

$$\alpha = \frac{1}{2}; \quad \varphi = e^1$$

$$y_{Ae} = 0.535$$

$$X_A = 1 - y_{Ae} = 0.465$$

(c) $y_{Ae} = \dfrac{(0.4)(\varphi - 1) + (1 - \alpha)}{\varphi - \alpha}$

As $\alpha \to 1$, $\phi = e^{2(1-\alpha)} \approx 1 + 2(1 - \alpha)$

$$y_{Ae} = \frac{(0.4)(2)(1 - \alpha) + (1 - \alpha)}{1 + 2(1 - \alpha) - \alpha}$$

$$= \frac{1.8(1 - \alpha)}{3(1 - \alpha)} = \frac{1.8}{3} = 0.6$$

But total moles of A reacted, $n_{A0}X_A = n_{A0}(y_{Ai} - y_{Ae})$

$$X_A = 0.4$$

Solutions to Chapter 4

4.1 $A \to 2R$ $r_{A1} = k_1 C_A$

       $A \to 3S$ $r_{A2} = k_2 C_A$

$n_R/n_S = 4$; $\tau = 40$ s; $x_A = 0.6$; $v_T = $ const; $n_{R0} = n_{S0} = 0$

$$\frac{r_{A1}}{r_{A2}} = \frac{k_1}{k_2}\frac{n_R/2}{n_S/3} = \frac{3n_R}{2n_S}$$

$$\frac{k_1}{k_2} = 6$$

$$V = -\int_{n_{A0}}^{n_A} \frac{dn_A}{(k_1 + k_2)n_A/v_T}$$

$$\tau = \frac{1}{k_1 + k_2} \ln \frac{1}{1 - x_A}$$

$$k_1 + k_2 = 0.0229 \text{ s}^{-1}$$

$$k_2 = \frac{0.0229}{7} = 0.00327 \text{ s}^{-1}$$

$$k_1 = 0.0196 \text{ s}^{-1}$$

4.2 $A \rightarrow 2R \quad r_{A1} = k_1 P_A$

$A \rightarrow 3S \quad r_{A2} = k_2 P_A$

$$\frac{n_{A1}}{n_{A2}} = \frac{k_1}{k_2}$$

$$n_A = n_{A0} - n_{A1} - n_{A2}$$

$$n_R = 2n_{A1}$$

$$n_S = 3n_{A2}$$

$$n_T = n_{A0} + n_{A1} + 2n_{A2}$$

$$Y_{R/A} = \frac{n_{A1}}{n_{A0}}, \quad Y_{S/A} = \frac{n_{A2}}{n_{A0}}, \quad x_A = Y_{R/A} + Y_{S/A}$$

$$n_T = n_{A0} + n_{A1} + 2\frac{k_2}{k_1}n_{A1} \quad \text{and} \quad n_A = n_{A0} - n_{A1} - \frac{k_2}{k_1}n_{A1}$$

$$n_{A1} = \frac{k_1}{k_1 + k_2}(n_{A0} - n_A)$$

$$n_T = n_{A0} + \frac{2k_2 + k_1}{k_2 + k_1}(n_{A0} - n_A)$$

Let $a = \dfrac{2k_2 + k_1}{k_2 + k_1} = \dfrac{5}{3}$

$$V = -\int_{n_{A0}}^{n_A} \frac{dn_A}{(k_1 + k_2)\frac{n_A}{n_T}P}$$

$$= -\frac{1}{P(k_1 + k_2)}\int_{n_{A0}}^{n_A} \frac{n_{A0} + a(n_{A0} - n_A)}{n_A}dn_A$$

$$= -\frac{1}{P(k_1 + k_2)}[(1+a)n_{A0} \ln n_A - an_A]_{n_{A0}}^{n_A}$$

$$= -\frac{1}{P(k_1 + k_2)}\left((1+a)n_{A0} \ln \frac{n_A}{n_{A0}} - an_A + an_{A0}\right)$$

$$= \frac{n_{A0}}{P(k_1 + k_2)}\left((1+a)\ln \frac{1}{1-x_A} - ax_A\right)$$

At exit $Y_{R/A} = 0.3$, $\dfrac{Y_{R/A}}{Y_{S/A}} = \dfrac{k_1}{k_2}$, $Y_{S/A} = 0.6$, $x_A = 0.9$

$$V = -\frac{1}{1(60)}\left[\frac{8}{3}\ln\frac{1}{0.1} - \frac{5}{3}(0.9)\right] = 0.0773 \text{ m}^3$$

4.3

(a) $r_A = -\dfrac{dn_A}{dV}$

$$V = -\int \frac{dn_A}{k_1 C_A + k_2 C_A^2} = -v_T \int \frac{dC_A}{k_1 C_A + k_2 C_A^2}$$

$$= -v_T \int \frac{dC_A}{k_1 C_A} - \frac{k_2}{k_1^2}\frac{dC_A}{\left(1 + \frac{k_2}{k_1}C_A\right)}$$

$$= \frac{v_T}{k_1}\left[\ln\frac{\left(1 + \frac{k_2}{k_1}C_A\right)}{C_A}\right]_{C_{A0}}^{C_A}$$

$$= \frac{v_T}{k_1}\ln\frac{1 + 10C_A}{C_A}\frac{C_{A0}}{1 + 10C_{A0}}$$

$$= \frac{(1)}{(1)}\ln\frac{(15)}{(0.05)}\frac{(1)}{(11)} = 1.00 \text{ litres}$$

(b) $r_B = \dfrac{dn_B}{dV} = k_1 C_A$

$$-\frac{dn_B}{dn_A} = -\frac{dC_B}{dC_A} = \frac{r_B}{r_A} = \frac{k_1 C_A}{k_1 C_A + k_2 C_A^2} = \frac{k_1}{k_1 + k_2 C_A}$$

$$C_B - C_{B0} = \int_{C_{A0}}^{C_A} \frac{-k_1}{k_1 + k_2 C_A}dC_A$$

$$= \left[-\frac{k_1}{k_2}\ln(k_1 + k_2 C_A)\right]_{C_{A0}}^{C_A}$$

$$= -\frac{1}{10}\ln\frac{11}{1.5} = 0.199 \text{ mol}^{-1}$$

$$Y_{B/A} = \frac{C_B - C_{B0}}{C_{A0}} = 19.9\%$$

(c) $S_{B/A} = \dfrac{C_B - C_{B0}}{C_{A0} - C_A} = \dfrac{0.199}{0.95} = 20.9\%$

(d) $V = \dfrac{n_{A0} - n_A}{r_A}$

$(C_{A0} - C_A)v_T = V(k_1 C_A + k_2 C_A^2)$

$\dfrac{C_{A0} V k_2}{v_T}\left(\dfrac{C_A}{C_{A0}}\right)^2 + \left(\dfrac{V k_1}{v_T} + 1\right)\dfrac{C_A}{C_{A0}} - 1 = 0$

$10\left(\dfrac{C_A}{C_{A0}}\right)^2 + 2\dfrac{C_A}{C_{A0}} - 1 = 0$

$\dfrac{C_A}{C_{A0}} = 0.232$

$x_A = 76.8\%$

We can see that the conversion of A is much lower in the CSTR.

(e) $\dfrac{n_B - n_{B0}}{V} = k_1 C_A$

$C_B - C_{B0} = k_1 C_A \dfrac{V}{v_T}$

$\qquad\qquad = (1)(0.232)\dfrac{(1.00)}{(1)} = 0.232 \text{ mol l}^{-1}$

$Y_{B/A} = \dfrac{C_B - C_{B0}}{C_{A0}} = 23.2\%$

However, the yield of B is greater in the CSTR than in the PFR despite the lower conversion. This can be explained by the more favourable selectivity obtained at low reactant concentrations (the favoured reaction is of lower order than the competing reaction).

(f) $S_{B/A} = \dfrac{C_B - C_{B0}}{C_{A0} - C_A} = \dfrac{0.232}{0.768} = 30.2\%$

(g) $C_{A0} x_A v_T = V(k_1 C_A + k_2 C_A^2)$

$(1)(0.95)(1) = V((1)(0.005) + (10)(0.005)^2)$

$V = 12.7 \text{ litres}$

$\dfrac{C_B - C_{B0}}{C_{A0}} = k_1 \dfrac{C_A}{C_{A0}}\dfrac{V}{v_T} = (0.05)(12.7) = 63.5\%$

(h) $S_{B/A} = \dfrac{0.635}{0.95} = 66.8\%$

4.4

(a)   Some B must be added to the reactor to start reaction 1. Once the reaction has started then the B produced will catalyse the reaction and the reaction will continue.

(b)   Design equation for reaction 1,

$$V = \frac{n_B}{k_1 C_A C_B}$$

$$C_A = \frac{1}{k_1 \tau}$$

Design equation for reaction 2,

$$V = \frac{n_C}{k_2 C_A}$$

$$C_C = k_2 C_A \tau = \frac{k_2}{k_1}$$

$$C_B = C_{A0} - C_A - C_C = C_{A0} - \frac{1}{k_1 \tau} - \frac{k_2}{k_1}$$

(c)   Must maximize $C_A C_B$ or $C_{A0}(1 - x_A)\left(C_{A0} - \frac{1}{k_1 \tau} - \frac{k_2}{k_1}\right)$

$$C_{A0}(1 - x_A)\left(C_{A0} - \frac{1}{k_1 \tau} - \frac{k_2}{k_1}\right) = C_{A0}(1 - x_A)\left(C_{A0} x_A - \frac{k_2}{k_1}\right)$$

$$C_{A0} x_A - \frac{k_2}{k_1} - C_{A0} x_A^2 + \frac{k_2}{k_1} x_A \text{ must be maximized,}$$

Differentiating the above expression with respect to x_A and setting equal to zero,

$$C_{A0} - 2 C_{A0} x_A + \frac{k_2}{k_1} = 0$$

$$x_A = \frac{1}{2}\left(1 + \frac{k_2}{k_1 C_{A0}}\right)$$

$$S_{B/A} = \frac{C_B}{C_{A0} - C_A} = \frac{C_B}{C_{A0}} \frac{C_{A0}}{C_{A0} - C_A} = \frac{\left(x_A - \frac{k_2}{k_1 C_{A0}}\right)}{x_A}$$

$$S_{B/A} = \frac{1 - \frac{k_2}{k_1 C_{A0}}}{1 + \frac{k_2}{k_1 C_{A0}}}$$

4.5 $A \rightarrow R \rightarrow S$

PFR: n_R is maximum when (eqn 4.46),

$$\tau = \frac{\ln k_2/k_1}{k_2 - k_1} = \frac{\ln 4}{1.5} = 0.924 \text{ s}$$

$$x_A = 1 - \frac{n_A}{n_{A0}} = 1 - e^{-k_1 \tau} = 0.843$$

$$n_R/n_{A0} = \frac{k_1}{k_2 - k_1}\{e^{-k_1\tau} - e^{-k_2\tau}\}$$

$$= -\frac{2}{1.5}(0.157 - 0.630) = 0.631$$

$$x_A = n_R/n_{A0} + n_S/n_{A0}$$

$$n_S/n_{A0} = 0.212$$

$$\text{CSTR}: \quad n_R/n_{A0} = \frac{k_1\tau}{(1 + k_1\tau)(1 + k_2\tau)}$$

$$\frac{dn_R}{d\tau} = 0$$

$$\frac{1}{n_{A0}} = \frac{dn_R}{d\tau} = \frac{-k_1\tau[(1 + k_1\tau)k_2 + (1 + k_2\tau)k_1] + (1 + k_1\tau)(1 + k_2\tau)k_1}{(1 + k_1\tau)^2(1 + k_2\tau)^2}$$

$$k_2\tau + k_1k_2\tau^2 + k_1\tau + k_1k_2\tau^2 - 1 - k_1\tau - k_2\tau - k_1k_2\tau^2 = 0$$
$$k_1k_2\tau^2 = 1$$

$$\tau = \frac{1}{\sqrt{k_1k_2}} = 1.00 \text{ s}$$

$$x_A = \frac{k_1\tau}{1 + k_1\tau} = 0.667$$

$$\frac{n_R}{n_{A0}} = \frac{k_1\tau}{(1 + k_1\tau) + (1 + k_2\tau)} = 0.444$$

$$n_S/n_{A0} = 0.223$$

4.6 $k_3/k_1 = 0.4$; $(k_2 + k_4)/k_1 = 0.2$

No B present initially, v_T is constant.

(a)   PFR : $n_A = n_{A0}\,e^{-(k_1 + k_3)\tau}$

$$\frac{n_A}{n_{A0}} = 1 - x_A = 0.3 = e^{-(k_1 + k_3)\tau} = e^{-1.4k_1\tau}$$

but $\dfrac{dn_B}{dV} = k_1C_A - (k_2 + k_4)C_B$

$$\frac{dn_B}{d\tau} = k_1n_A - (k_2 + k_4)n_B$$

$$\frac{dn_B}{d\tau} = (k_2 + k_4)n_B = k_1n_{A0}e^{-(k_1 + k_3)\tau}$$

Solution of the form:

$$n_B = Ae^{-(k_2 + k_4)\tau} + Be^{-(k_1 + k_3)\tau}$$

$$\frac{dn_B}{d\tau} = -(k_2 + k_4)Ae^{-(k_2+k_4)\tau} - (k_1 + k_3)Be^{-(k_1+k_3)\tau}$$

As $n_B = 0$ when $\tau = 0$: $A + B = 0$

Coefficients of $e^{-(k_1+k_3)\tau}$:

$$-(k_1 + k_3)B + (k_2 + k_4)B = k_1 n_{A0}$$

$$B = \frac{k_1 n_{A0}}{(k_2 + k_4) - (k_1 + k_3)}; \quad A = -B$$

$$n_B = \frac{k_1 n_{A0}}{(k_2 + k_4) - (k_1 + k_3)} \{e^{-(k_1+k_3)\tau} - e^{-(k_2+k_4)\tau}\}$$

$$= \frac{k_1 n_{A0}}{(k_1 + k_3) - (k_2 + k_4)} \{e^{-(k_2+k_4)\tau} - e^{-(k_1+k_3)\tau}\}$$

$$e^{-(k_1+k_3)\tau} = e^{-1.4k_1\tau} = 0.3$$

$$e^{-(k_2+k_4)\tau} = e^{-0.2k_1\tau} = (0.3)^{1/7}$$

$$n_B = \frac{n_{A0}}{1.4 - 0.2} \{(0.3)^{1/7} - 0.3\}$$

$$\frac{n_B}{n_{A0}} = 0.452$$

(b)   CSTR : $\dfrac{n_{A0} - n_A}{V} = (k_1 + k_3)\dfrac{n_A}{V_T}$

$$n_A = \frac{n_{A0}}{1 + (k_1 + k_3)\tau}$$

$$\frac{n_B - n_{B0}}{V} = k_1 \frac{n_A}{v_T} - (k_2 + k_4)\frac{n_B}{v_T}$$

$$n_B = \frac{k_1 \tau n_{A0}}{1 + (k_1 + k_3)\tau} - (k_2 + k_4)\tau n_B$$

$$= \frac{k_1 \tau n_{A0}}{\{1 + (k_1 + k_3)\tau\}\{1 + (k_2 + k_4)\tau\}}$$

but $x_A = 1 - \dfrac{n_A}{n_{A0}} = 1 - \dfrac{1}{1 + (k_1 + k_3)\tau}$

$$= \frac{(k_1 + k_3)\tau}{1 + (k_1 + k_3)\tau} = 0.7$$

$(k_1 + k_3)\tau = 2.333$

$k_1\tau = 1.667$

$(k_2 + k_4)\tau = 0.333$

$$n_B/n_{A0} = \frac{1.667}{\{1 + 2.333\} + \{1 + 0.333\}} = 0.375$$

Solutions to Chapter 5

5.1

(a) $\quad V = \dfrac{n_{A0} - n_A}{r_A} = \dfrac{n_{A0}x_A}{k_1 C_A C_B}$

$$C_A = C_B = \frac{n_A}{v_T} = C_{A0}(1 - x_A)$$

$$V = \frac{n_{A0}x_A}{k_1 C_{A0}^2 (1 - x_A)^2}; \; n_{A0} = C_{A0}v_T$$

$$k_1 = \frac{x_A v_T}{V C_{A0}(1 - x_A)^2}$$

$$= \frac{(0.6)(1)}{(4)(3)(0.4)^2} \frac{\text{litre s}^{-1}}{\text{litre mol litre}^{-1}} = 0.3125 \text{ litre mol}^{-1} \text{ s}^{-1}$$

$$k_1 = 10^{14} e^{-\Delta E/RT}$$

$$T = -\frac{\Delta E}{R} \Big/ \ln \frac{k_1}{10^{14}}$$

$$= \frac{-10^5/8.314}{\ln(0.3125 \times 10^{-14})} = 360.1 \text{ K}$$

(b) Energy balance:

$$v_T \bar{c}_P (T_0 - T) - \Delta H_R x_A n_{A0} + Q = 0$$

$$Q = (1)(4.2)(70.1) + (60)(0.6)(3) \text{ kW} = 402 \text{ kW (added)}$$

5.2 $\quad A \Leftrightarrow B; \quad r_A = k_1 P_A - k_1 P_B$

$$P_A = \frac{n_{A0}}{n_T}(1 - x_A)P; \; P_B = \frac{n_{A0}x_A}{n_T} \cdot P$$

$$r_A \frac{n_T}{n_{A0}} \frac{1}{P} = k_1 - (k_1 + k_{-1})x_A$$

$$\frac{n_T}{n_{A0}P} \frac{dr_A}{dT} = \frac{dk_1}{dT} - \left(\frac{dk_1}{dT} + \frac{dk_{-1}}{dT}\right)x_A$$

and $\frac{dr_A}{dT} = 0$ for maximum rate at a given conversion

$$x_A = \frac{dk_1/dT}{dk_1/dT + dk_{-1}/dT}$$

$$= \frac{1}{1 + \frac{(\frac{-\Delta E_{-1}}{R})(-1/T^2)k_{-1}'e^{-\Delta E_{-1}/RT}}{(\frac{-\Delta E_1}{R})(-1/T^2)k_1'e^{-\Delta E_1/RT}}}$$

$$= \frac{1}{1 + \frac{\Delta E_{-1}}{\Delta E_1} \cdot \frac{k_{-1}}{k_1}}$$

$$= \frac{1}{1 + (2 \times 10^3 e^{-5000/T})}$$

$$T = \frac{-5000}{\ln\left\{\frac{\frac{1}{0.62} - 1}{2000}\right\}} = 618 \text{ K}$$

At this temperature

$$r_A = \frac{n_{A0}P}{n_T}\{k_1 - (k_1 + k_{-1})x_A\}$$

$$k_1 = 0.306 \text{ mol s}^{-1} \text{ m}^{-3} \text{ bar}^{-1}$$

$$k_{-1} = 0.094 \text{ mol s}^{-1} \text{ m}^{-3} \text{ bar}^{-1}$$

$$r_A = \left(\frac{1}{10}\right)(1)\{0.306 - (0.4)(0.62)\}\text{mol s}^{-1} \text{ m}^{-3} = 5.8 \times 10^{-3} \text{ mol s}^{-1} \text{ m}^{-3}$$

$$V = \frac{n_{A0}x_A}{r_A} = \frac{(0.21)(0.62)}{(5.8 \times 10^{-3})} \text{ m}^3 = 22.4 \text{ m}^3$$

Energy balance:

$$n_T c_P (T - T_0) = (-\Delta H_R)x_A n_{A0} + Q$$

$$\Delta H_R = \Delta E_1 - \Delta E_2 = -41570 \text{ J mol}^{-1}$$

$$Q = (2.1)(42)(45) - (41570)(0.62)(0.21) \text{ W} = -1443 \text{ W}$$

5.3

(a) Reactor 2 operates at 618 K to maximize rate (see solution for Problem 5.2),

$$V_2 = \frac{n_{A0}(x_{A2} - x_{A1})}{r_{A2}}$$

$$x_{A2} - x_{A1} = \frac{(5)(5.8 \times 10^{-3})}{(0.21)} = 0.138$$

$$x_{A1} = 0.482$$

$$T_1 = \frac{-5000}{\ln\left\{\frac{\frac{1}{0.482}-1}{2000}\right\}} = 664 \text{ K}$$

At this temperature: $k_1 = 0.537 \text{ mol s}^{-1} \text{ m}^{-3} \text{ bar}^{-1}$

$$k_{-1} = 0.288 \text{ mol s}^{-1} \text{ m}^{-3} \text{ bar}^{-1}$$

$$V_1 = \frac{(0.21)(0.482)}{(0.1)\{0.537 - (0.537 + 0.288)(0.482)\}} \text{ m}^3 = 7.40 \text{ m}^3$$

(b) Adiabatic temperature rises:

$$n_T \overline{c_P} \Delta T = (-\Delta H_R)n_{A0}\Delta x_A$$

$$\Delta T_1 = \frac{(41570)}{(10)}\frac{(0.482)}{(42)} = 47.7 \text{ K}, \, T_1(\text{feed}) = 664 - 44.7 = 616 \text{ K}$$

$$\Delta T_2 = \frac{41570}{10}\frac{(0.138)}{42} = 13.7 \text{ K}, \, T_2(\text{feed}) = 618 - 13.7 = 604 \text{ K}$$

(c) With 5 m³ reactor first, to maximize rate in reactor 1:

$$T_1 = \frac{-5000}{\ln\left\{\frac{\frac{1}{x_{A1}}-1}{2000}\right\}}$$

$$V_1 = \frac{n_{A0}x_{A1}}{\frac{n_{A0}P}{n_T}\{k_1 - (k_1 + k_{-1})x_{A1}\}}$$

$$g(x_A, T) = 1 = \frac{n_T x_{A1}}{PV_1\{k_1 - (k_1 + k_{-1})x_{A1}\}}$$

$$\frac{n_T}{PV_1} = 0.42 \text{ mol s}^{-1}\text{ m}^{-3}\text{ bar}^{-1}$$

$$g(x_A, T) = \frac{0.42\, x_{A1}}{k_1 - (k_1 + k_{-1})x_{A1}}$$

We guess x_{A1} and evaluate the temperature, the rate constants, and g,

$$x_{A1} = 0.435$$

Reactor 2 operates at 618 K again.

$$V_2 = \frac{n_{A0}(x_{A2} - x_{A1})}{r_{A2}} = \frac{(0.21)(0.185)}{(5.8 \times 10^{-3})}\text{ m}^3 = 6.70 \text{ m}^3$$

x_{A1}	$k_1/\text{mol s}^{-1}$ $\text{m}^{-3}\text{ bar}^{-1}$	$g(x_A, T)$
0.43	0.663	0.956
0.435	0.649	0.996
0.44	0.636	1.037

5.4 $A \rightarrow B$ $\quad r_A = k_1 P_A$

Energy balance (adiabatic PFR):

$$\frac{dT}{dV} = -\frac{r_A \Delta H_R}{n_{T0}c_P}$$

Material balance:

$$\frac{dx_A}{dV} = \frac{r_A}{n_{A0}}$$

$$\frac{dT}{dx_A} = \frac{n_{A0}}{n_{T0}}\frac{(-\Delta H_R)}{\overline{c_P}} = \frac{1}{2}\left(\frac{5 \times 10^4}{100}\right)$$

$$\frac{dT}{dx_A} = 250, T - T_0 = 250x_A$$

$$r_A = k_1 P\frac{n_A}{n_T} = k_1 P\frac{n_{A0}}{n_T}(1 - x_A)$$

$$k_1 = k_1' \exp\left(\frac{-\Delta E}{RT}\right)$$

$$V = \int_0^{x_A} \frac{n_{A0}dx_A}{P\frac{n_{A0}}{n_T}(1 - x_A)k_1'e^{-\Delta E/RT}}$$

$$= \frac{n_T}{Pk_1'}\int_0^{x_A} \frac{dx_A}{(1 - x_A)e^{-\Delta E/R(T_0 + 250x_A)}}$$

$$= 0.224 \int_0^{0.8} \frac{dx_A}{(1 - x_A) \exp\left\{\frac{-1000}{3000 + 250x_A}\right\}}$$

$$f(x_A) = \frac{0.224}{(1 - x_A) \exp\left\{\frac{-1000}{3000 + 250x_A}\right\}}$$

x_A	T	k_1	$f(x_A)/m^3$
0	300	1.00	6.285
0.1	325	1.30	5.404
0.2	350	1.61	4.880
0.3	375	1.95	4.610
0.4	400	2.31	4.552
0.5	425	2.67	4.716
0.6	450	3.05	5.172
0.7	475	3.42	6.135
0.8	500	3.80	8.283

Trapezoidal rule:

$$\text{Integral} = \frac{(f_0 + f_1)}{2} \cdot \Delta x_A + \frac{(f_1 + f_2)}{2} \Delta x_A + \ldots + \frac{(f_7 + f_8)}{2} \Delta x_A$$

$$V \approx 4.3 \ \text{m}^3$$

5.5　$r_A = -\dfrac{1}{V} \dfrac{dN_A}{dt}$

$$\int_0^t dt = N_{A0} \int_0^{x_A} \frac{1}{V} \frac{dx_A}{kP_A}$$

$$P_A = \frac{N_A}{N_T} P, \ PV = \frac{N_T RT}{10^5} \quad (P \ \text{in bar})$$

$$P_A = \frac{N_{A0}(1 - x_A)RT}{10^5 V}$$

$$\int_0^t dt = 10^5 \int_0^{x_A} \frac{dx_A}{k(1 - x_A)RT}$$

$$\int_0^t kRT dt = 10^5 \int_0^{x_A} \frac{dx_A}{(1 - x_A)}$$

$$T = 300 + 0.2t$$

Let $\displaystyle\int_0^t kRT dt = \int_0^t f(t)dt$

$$\int_0^{1200} f(t)dt = 200\left(\frac{4.02}{2} + 11.72 + \ldots 165.8 + \frac{255.7}{2}\right) = 0.982 \times 10^5$$

$$0.982 = [-\ln(1 - x_A)]_0^{x_A}$$

$$\ln\frac{1}{1 - x_A} = 0.982$$

$$x_A = 0.62$$

Note: this question does not actually involve an energy balance and could have been solved using only the material from Chapter 3 and earlier.

t/s	T/K	f
0	300	4.02
200	340	11.72
400	380	27.57
600	420	55.69
800	460	100.4
1000	500	165.8
1200	540	255.7

5.6 $A \rightarrow B + C$

Energy balance:

$$n_{T0}\overline{c_{PA}}(T - T_0) + UA(T - T_j) = -\Delta H_R n_{A0} x_A$$

$$x_A = \frac{n_{T0}\overline{c_{PA}}}{-\Delta H_R n_{A0}}(T - T_0) + \frac{UA}{-\Delta H_R n_{A0}}(T - T_j)$$

$$= \frac{120}{7 \times 10^4}(T - 325) + \frac{5 \times 10^{-2}}{(7 \times 10^{-4}(10^{-3})}(T - 373.2)$$

$$= \frac{170}{7 \times 10^4}(T - 339.2)$$

$$n_A = n_{A0} - n_{A0}x_A, n_B = n_{A0}x_A, n_C = n_{A0}x_A, n_T = n_{A0} + n_{A0}x_A$$

$$V = \frac{n_{A0}x_A}{kn_{A/V_T}} = \frac{n_{A0}x_A}{kn_{A0}(1 - x_A)} \cdot n_{A0}(1 + x_A)\frac{RT}{P}$$

(As $v_T = n_T RT/P$ for a perfect gas)

$$1 = \frac{x_A(1 + x_A)}{(1 - x_A)} \cdot \frac{T}{k} \cdot \frac{n_{A0}R}{PV}$$

$$\frac{n_{A0}R}{PV} = 4.16 \times 10^{-6} \text{ s}^{-1} \text{ K}^{-1}$$

Let $f(x_A, T) = x_A \frac{x_A(1 + x_A)}{(1 - x_A)} \frac{T}{10^{11}e^{-18000/T}}(4.16 \times 10^{-6})$

Use iterative procedure—guess T, calculate x_A from energy balance, substitute in expression for $f(x_A, T)$ which should equal unity.

$T = 592 \text{ K}, x_A = 0.61$

This is the unstable operating point because there are two other solutions on either side,

$x_A \approx 0; T = 339.2 \text{ K}$, Therefore $x_A = 6.4 \times 10^{-10}$

$x_A \approx 1; T = 751 \text{ K}$, Therefore $1 - x_A \approx 1.6 \times 10^{-3}, x_A = 0.9984$

T/K	x_A	$f(x_A, T)$
550	0.512	5.93
592	0.614	1.013
595	0.621	0.904
600	0.633	0.751

$T = 592 \text{ K}, x_A = 0.61$

Nomenclature

Below is a list of common symbols used. Typical units are shown in parentheses; however, kJ, kmol, litres, and kW are used interchangeably with J, mol, m^3, and W.

A	Heat transfer area or cross-sectional area (m^2)
C_A	Concentration of species A (mol m^{-3})
$\overline{C_A}$	Mean concentration of species A (mol m^{-3})
c_P	Specific heat capacity (J mol^{-1} K^{-1} or J m^{-3} K^{-1})
$\overline{c_P}$	Mean specific heat capacity (J mol^{-1} K^{-1} or J m^{-3} K^{-1})
$E(t)$	Residence time distribution (s^{-1})
k	Reaction rate constant (units vary)
k'	Pre-exponential constant in expression for reaction rate constant (units vary)
K	Equilibrium constant (units vary)
l	Length variable (m)
L	Total length of reactor (m)
n	Reaction order (dimensionless)
n_A	Molar flow rate of species A (mol s^{-1})
n_T	Total molar flow rate (mol s^{-1})
N_A	Moles of A in a closed system (mol)
N_T	Total moles in a closed system (mol)
n_{A1}	Molar flow of A reacted or produced through reaction 1 (mol s^{-1})
N_{A1}	Moles of A reacted or produced through reaction 1 (mol)
P_A	Partial pressure of A (bar)
q	Heat flux (W m^{-2})
Q	Heat transferred from a system (W)
Q_g	Heat generated by reaction (W)
Q_r	Heat removed by cooling or heating of the process stream (W)
r_A	Rate of reaction or formation of A (mol m^{-3} s^{-1})
r_{Ai}	Rate of reaction or formation of A through reaction i (mol m^{-3} s^{-1})
R	Gas constant (= 8.314 J mol^{-1} K^{-1})
Re	Reynold's number
$S_{B/A}$	Overall selectivity for B with reference to A consumed (dimensionless)
S_V	Space velocity (s^{-1})
t	Time variable (s)
$\bar{t}$	mean residence time (s)
T	Temperature (K)
T_j, T_0	Temperature of cooling/heating jacket and feed respectively (K)
U	Overall heat transfer coefficient (W m^{-2} K^{-1})

V	Reactor volume variable for PFR, total volume for a batch reactor or CSTR (m^3)
v_T	Total volumetric flow rate (m^{-3} s^{-1})
x_A	Conversion of A or 'per-pass conversion' of A (dimensionless)
$x_{A,Ri}$	Conversion of A for an individual reactor i (dimensionless)
X_A	Overall conversion of A for a reactor with recycle (dimensionless)
y_A	Mole fraction of A (dimensionless)
$Y_{B/A}$	Yield of B in terms of A fed to the reactor (dimensionless)
α	Fraction recycled (dimensionless)
δ	Dirac delta function
$\phi_{B/A}$	Local selectivity for B with reference to A consumed (dimensionless)
ν	Stoichiometric coefficient (dimensionless)
θ	Total residence time of a batch reactor (s)
τ	Mean residence time of a continuous reactor, residence time of a PFR (s)
ξ	Extent of reaction (dimensionless)
ΔE	Activation energy (J mol^{-1})
ΔH_R	Heat of reaction (J mol^{-1})
ΔU	Change in internal energy (J mol^{-1})

Subscripts

i	Inlet conditions to a reactor with recycle; intermediate stream
i	ith reaction; ith species; ith reactor
e	Exit conditions
0	Inlet conditions or time $t = 0$

Superscripts

*	Denotes equilibrium

Further reading

This book is designed to serve as an introductory text to the subject of chemical reaction engineering. The following texts are suggested to broaden the reader's perspective and to develop more advanced aspects of the subject (reference [4], in particular, is more advanced).

[1] 'Chemical reaction engineering', Levenspiel, O., 2nd ed. (1972).
[2] 'Chemical reactor theory: an introduction', Denbigh, K. G. and Turner, J. C. R., 3rd ed. (1984).
[3] 'Elements of chemical reaction engineering', Fogler, H. S., 2nd ed. (1992).
[4] 'Chemical reactor analysis and design', Froment, G. F. and Bischoff, K. B., 2nd ed. (1990).

Index